Design, Simulation and Applications of Inductors and Transformers for Si RF ICs

THE KLUWER INTERNATIONAL SERIES IN ENGINEERING AND COMPUTER SCIENCE

ANALOG CIRCUITS AND SIGNAL PROCESSING
Consulting Editor: **Mohammed Ismail**. *Ohio State University*

Related Titles:

DESIGN AND IMPLEMENTATION
    B.E. Jonsson
    ISBN: 0=7923-7871-7
RESEARCH PERSPECTIVES ON DYNAMIC TRANSLINEAR AND LOG-DOMAIN CIRCUITS
    W.A. Serdijn, J. Mulder
    ISBN: 0-7923-7811-3
CMOS DATA CONVERTERS FOR COMMUNICATIONS
    M. Gustavsson, J. Wikner, N. Tan
    ISBN: 0-7923-7780-X
DESIGN AND ANALYSIS OF INTEGRATOR-BASED LOG -DOMAIN FILTER CIRCUITS
    G.W. Roberts, V. W. Leung
    ISBN: 0-7923-8699-X
VISION CHIP
    A. Moini
    ISBN: 0-7923-8664-7
COMPACT LOW-VOLTAGE AND HIGH-SPEED CMOS, BiCMOS AND BIPOLAR OPERATIONAL AMPLIFIERS
    K-J. de Langen, J. Huijsing
    ISBN: 0-7923-8623-X
CONTINUOUS-TIME DELTA-SIGMA MODULATORS FOR HIGH-SPEED A/D CONVERTERS: Theory, Practice and Fundamental Performance Limits
    J.A. Cherry, W. M. Snelgrove
    ISBN: 0-7923-8625-6
LEARNING ON SILICON: Adaptive VLSI Neural Systems
    G. Cauwenberghs, M.A. Bayoumi
    ISBN: 0=7923-8555-1
ANALOG LAYOUT GENERATION FOR PERFORMANCE AND MANUFACTURABILITY
    K. Larnpaert, G. Gielen, W. Sansen
    ISBN: 0-7923-8479-2
CMOS CURRENT AMPLIFIERS
    G. Palmisano, G. Palumbo, S. Pennisi
    ISBN: 0-7923-8469-5
HIGHLY LINEAR INTEGRATED WIDEBAND AMPLIFIERS: Design and Analysis Techniques for Frequencies from Audio to RF
    H. Sjöland
    ISBN: 0-7923-8407-5
DESIGN OF LOW-VOLTAGE LOW-POWER CMOS DELTA-SIGMA A/D CONVERTERS
    V. Peluso, M. Steyaert. W. Sansen
    ISBN: 0-7923-8417-2
THE DESIGN OF LOW-VOLTAGE, LOW-POWER SIGMA-DELTA MODULATORS
    S. Rabii, B.A. Wooley
    ISBN: 0-7923-8361-3
TOP-DOWN DESIGN OF HIGH-PERFORMANCE SIGMA-DELTA MODULATORS
    F. Medeiro, A. Pérez-Verdú, A. Rodríguez-Vázquez
    ISBN: 0-7923-8352-4

DESIGN, SIMULATION AND APPLICATIONS OF INDUCTORS AND TRANSFORMERS FOR SI RF ICS

ALI M. NIKNEJAD
Silicon Laboratories, Austin TX

ROBERT G. MEYER
University of California, Berkeley

Kluwer Academic Publishers
Boston/Dordrecht/London

Distributors for North, Central and South America:
Kluwer Academic Publishers
101 Philip Drive
Assinippi Park
Norwell, Massachusetts 02061 USA
Telephone (781) 871-6600
Fax (781) 871-6528
E-Mail <kluwer@wkap.com>

Distributors for all other countries:
Kluwer Academic Publishers Group
Distribution Centre
Post Office Box 322
3300 AH Dordrecht, THE NETHERLANDS
Telephone 31 78 6392 392
Fax 31 78 6546 474
E-Mail <orderdept@wkap.nl>
 Electronic Services <http://www.wkap.nl>

Library of Congress Cataloging-in-Publication Data

Niknejad, Ali M., 1972-
 Design, simulation and applications of inductors and transformers
 for Si RF ICs / Ali M. Niknejad, Robert G. Meyer.
     p. cm. -- (Kluwer international series in engineering and
     computer science. Analog circuits and signal processing)
 Includes bibliographical references and index.
 ISBN 0-7923-7986-1 (alk. paper)
 1. Electric inductors--Computer-aided design. 2. Electronic
transformers--Computer-aided design. 3. Silicon-on-insulator
technology. 4. Microwave integrated circuits--Computer-aided design.
 I. Meyer, Robert G., 1942- II. Title. III. Series

TK7872.I63 N55 2000
621.381--dc21
                                                           00-062480
                                                              CIP

Copyright © 2000 by Kluwer Academic Publishers..

All rights reserved. No part of this publication may be reproduced, stored in a retrieval system or transmitted in any form or by any means, mechanical, photo-copying, recording, or otherwise, without the prior written permission of the publisher, Kluwer Academic Publishers, 101 Philip Drive, Assinippi Park, Norwell, Massachusetts 02061

Printed on acid-free paper.
Printed in the United States of America

Contents

List of Figures	ix
List of Tables	xv
Preface	xvii
Acknowledgments	xix

Part I Analysis and Simulation of Passive Devices

1. INTRODUCTION		3
1.	Introduction	3
2.	Passive Devices in Early Integrated Circuits	3
3.	Applications of Passive Devices	4
4.	Wireless Communication	6
5.	Si Integrated Circuit Technology	8
6.	Contributions of this Research	9
2. PROBLEM DESCRIPTION		11
1.	Definition of Passive Devices	11
2.	Loss Mechanisms	15
3.	Device Layout	21
4.	Substrate Coupling	35
3. PREVIOUS WORK		39
1.	Early Work	39
2.	Passive Devices on the GaAs substrate	40
3.	Passive Devices on the Si substrate	40
4.	Passive Devices on Highly Conductive Si Substrate	43
4. ELECTROMAGNETIC FORMULATION		45
1.	Introduction	45
2.	Maxwell's Equations	45
3.	Calculating Substrate Induced Losses	48

	4. Inversion of Maxwell's Differential Equations	50
	5. Numerical Solutions of Electromagnetic Fields	52
	6. Discretization of Maxwell's Equations	53

5. INDUCTANCE CALCULATIONS — 59
1. Introduction — 59
2. Definition of Inductance — 60
3. Parallel and Series Inductors — 65
4. Filamental Inductance Formulae for Common Configurations — 66
5. Calculation of Self and Mutual Inductance for Conductors — 67
6. High Frequency Inductance Calculation — 68

6. CALCULATION OF EDDY CURRENT LOSSES — 75
1. Introduction — 75
2. Electromagnetic Formulation — 76
3. Eddy Current Losses at Low Frequency — 84
4. Eddy Currents at High Frequency — 88
5. Examples — 93

7. ASITIC — 97
1. Introduction — 97
2. ASITIC Organization — 99
3. Numerical Calculations — 100
4. Circuit Analysis — 101

8. EXPERIMENTAL STUDY — 109
1. Measurement Results — 109
2. Device Calibration — 110
3. Single Layer Inductor — 113
4. Multi-Layer Inductor — 119

Part II   Applications of Passive Devices

9. VOLTAGE CONTROLLED OSCILLATORS — 125
1. Introduction — 125
2. Motivation — 127
3. Passive Device Design and Optimization — 128
4. VCO Circuit Design — 134
5. VCO Implementation — 141
6. Measurements — 144
7. Conclusion — 146

10. DISTRIBUTED AMPLIFIERS — 149
1. Introduction — 149
2. Image Parameter Method — 150

3.	Distributed Amplifier Gain	156
11. CONCLUSION		163
1.	Future Research	164

Appendix
A– Distributed Capacitance   167

List of Figures

1.1	Applications of passive devices in Si IC building blocks. (a) Impedance matching. (b) Tuned load. (c) Emitter degeneration. (d) Filtering. (e) Balun. (f) Distributed amplifier.	4
1.2	Traditional superheterodyne transceiver architecture.	6
1.3	Zero-IF direct conversion receiver architecture transceiver.	7
2.1	(a) An arbitrary black box with externally accessible ports. (b) The contents of two example "black boxes".	11
2.2	Various loss mechanisms present in an IC process.	15
2.3	Cross-section of metal and polysilicon layers in a typical IC process.	16
2.4	Schematic representation of substrate currents. Eddy currents are represented by the dashed lines and electrically induced currents by the solid lines.	19
2.5	Cross-section of typical (a) bipolar and (b) CMOS substrate layers.	20
2.6	The typical coil inductor.	22
2.7	A circular spiral inductor.	23
2.8	A square spiral inductor.	23
2.9	A polygon spiral inductor.	24
2.10	A symmetric spiral inductor.	24
2.11	A symmetric polygon spiral inductor.	25
2.12	A tapered spiral inductor.	27
2.13	A planar square spiral transformer.	28
2.14	An expanded view of a non-planar symmetric spiral transformer. The primary is shown on the left and the secondary on the right. These inductors actually reside on top of one another.	29

2.15	A planar symmetric balun transformer.	29
2.16	A patterned ground shield.	30
2.17	Substrate coupling in a mixed-signal system on a chip.	35
5.1	The reactance of a lossless passive device as a function of frequency.	61
5.2	(a) An isolated current loop. (b) A magnetically coupled pair of loops. Current only flows in loop j.	61
5.3	(a) Ground current at low frequency. (b) Ground current at high frequency.	68
5.4	The cross-sectional area of a return path loop (a) far removed from the conductor and (b) near the conductor.	69
5.5	The self and mutual inductance of the conductor and return path.	69
5.6	Two inductors realized with different metal spacing values of 1 μm and 10 μm.	70
5.7	The ratio of AC to DC inductance and resistance as a function of frequency plotted for two inductors of Fig. 5.6.	71
5.8	The current density at (a) 1 GHz and (b) 5 GHz.	71
5.9	The current density at (a) 1 GHz and (b) 5 GHz.	71
6.1	Multi-layer substrate excited by a filamental current source.	77
6.2	Single layer semi-infinite substrate excited by a filamental current source.	82
6.3	Cross-section of square spiral inductor.	85
6.4	Voltages and currents along series-connected 2-port elements.	89
7.1	*ASITIC* users can move between geometric (layout), electrical (inductance, Q, self-resonance), physical (technology file), and network (two-port parameters) domains.	98
7.2	A block diagram of the *ASITIC* modules.	99
7.3	An electrically short segment of the device.	101
7.4	(a) A grounded transformer. (b) The primary and secondary connected back-to-back with center point grounded.	104
7.5	A three-port balun.	105
8.1	Inductor test structures.	110
8.2	Layout of inductor L27.	112
8.3	Layout of inductor L19. In part (b) the lower metal layers have been staggered to give a clear picture of the device geometry.	112
8.4	Calibration pad structures. (a) S-G structure. (b) G-S-G structure.	113

List of Figures   xi

8.5	Measured and simulated (a) magnitude and (b) phase s-parameters of spiral inductor L27.	114
8.6	Measured and simulated inductance (imaginary component of Y_{21}) of spiral inductor L27.	115
8.7	Measured and simulated resistance (real component of Y_{21}) of spiral inductor L27.	115
8.8	Measured and simulated quality factor (imaginary over real component of Y_{21}) of spiral inductor L27.	116
8.9	Measured and simulated substrate resistance of spiral inductor L27.	117
8.10	Measured and simulated substrate capacitance of spiral inductor L27.	117
8.11	Measured and simulated (a) magnitude and (b) phase s-parameters of spiral inductor L19.	118
8.12	Measured and simulated inductance (imaginary component of Y_{21}) of spiral inductor L19.	119
8.13	Measured and simulated resistance (real component of Y_{21}) of spiral inductor L19.	120
8.14	Simple equivalent circuit of inductor L19 close to self-resonance.	120
8.15	(a) Measured and simulated quality factor (imaginary over real component of Y_{21}) of spiral inductor L19. (b) Close-up plot.	121
9.1	Simplified block diagram of a superheterodyne transceiver front-end.	125
9.2	The simulated quality factor of a small footprint 1 nH inductor versus an optimal 10 nH inductor. The smaller inductor dimensions are $R = 75\mu m$ $W = 5\mu m$ $N = 2$ while the larger inductor has dimensions $R = 150\mu m$ $W = 12.3\mu m$ $N = 7.5$. Both structures are realized with a metal spacing $S = 2.1\mu m$.	128
9.3	Layout of a center-tapped spiral inductor with radius $R = 125$ μm and the metal width $W = 14.5$ μm.	130
9.4	Simulated quality factor driven single-endedly (bottom curve) versus differentially (top curve). Note that the comparison is made with both a circular (higher Q value) and square device.	131
9.5	Compact circuit model for center-tapped spiral inductor.	131
9.6	Schematic of differential varactors.	132
9.7	Simulated varactor resonance curve.	132

xii   INDUCTORS AND TRANSFORMERS FOR SI RF ICS

9.8	Shielded MIM capacitor equivalent circuit. In (a) the top, bottom, and shielding plates are shown. Part (b) shows the equivalent circuit extracted from *ASITIC*. In (c) the load capacitor is shown realized as a symmetric-quad. In (d) the passive capacitive feedback network is shown, also realized as a pair of shielded symmetric quads.	133
9.9	Different feedback mechanisms for negative resistance generation.	135
9.10	Schematic of differential VCO.	136
9.11	Simplified equivalent circuit of differential VCO.	137
9.12	Simulated and calculated phase noise spectrum.	140
9.13	VCO core circuit schematic.	141
9.14	Different techniques to convert the VCO frequency: down-conversion, mode-locking, and latch-based frequency division.	142
9.15	Schematic of differential latch.	143
9.16	Schematic of output buffers.	144
9.17	Chip-level die photo of 2.9 GHz design utilizing a circular spiral.	144
9.18	Measurement setup.	145
9.19	Measured phase noise of VCO.	145
10.1	A three-stage distributed amplifier using CMOS n-FETs.	149
10.2	T-Section Network.	151
10.3	Lossless cascade of T-sections forming an artificial transmission line.	151
10.4	Lossy "gate" transmission line.	152
10.5	Lossy "drain" transmission line.	152
10.6	Gate and drain image impedance.	154
10.7	Gate line attenuation constant.	154
10.8	Drain line attenuation constant.	155
10.9	Bisected-π m-derived matching network.	155
10.10	Gate line input impedance using bisected-π m-derived matching network. Bottom curve shows the performance of a lossless transmission line. Top curve shows performance with lossy matching network. The flat line is the ideal gate impedance.	156
10.11	Simple FET model used for analysis.	156
10.12	Derivation of gate-voltage at the center of the T-section.	157

10.13	Gain as a function of frequency using Beyer's ideal expression (top curve) and the gain calculated in this work with an m-derived matching network (flat increasing curve) and without a matching network in place (rapidly decaying curve).	159
10.14	Gain versus the number of stages evaluated at low frequency (2 GHz).	160
10.15	Active load proposal to reduce gate or drain line losses.	161
10.16	Actively loaded gate attenuation constant in action. Propagation constant remains imaginary (not shown).	161
A.1	Schematic of a distributed lossy capacitor terminated in an arbitrary impedance.	167
A.2	A short segment of the distributed lossy capacitor.	168

List of Tables

8.1	CMOS Process Parameters Summary	111
8.2	Device Physical Dimensions	111
9.1	VCO circuit and process parameters.	140
9.2	Summary of measured performance.	146

Preface

The modern wireless communication industry has put great demands on circuit designers for smaller, cheaper transceivers in the gigahertz frequency range. One tool which has assisted designers in satisfying these requirements is the use of on-chip inductive elements (inductors and transformers) in silicon (Si) radio-frequency (RF) integrated circuits (ICs). These elements allow greatly improved levels of performance in Si monolithic low-noise amplifiers, power amplifiers, up-conversion and down-conversion mixers and local oscillators. Inductors can be used to improve the intermodulation distortion performance and noise figure of small-signal amplifiers and mixers. In addition, the gain of amplifier stages can be enhanced and the realization of low-cost on-chip local oscillators with good phase noise characteristics is made feasible.

In order to reap these benefits, it is essential that the IC designer be able to predict and optimize the characteristics of on-chip inductive elements. Accurate knowledge of inductance values, quality factor (Q) and the influence of adjacent elements (on-chip proximity effects) and substrate losses is essential. In this book the analysis, modeling and application of on-chip inductive elements is considered. Using analyses based on Maxwells equations, an accurate and efficient technique is developed to model these elements over a wide frequency range. Energy loss to the conductive substrate is modeled through several mechanisms, including electrically induced displacement and conductive currents and by magnetically induced eddy currents. These techniques have been compiled in a user-friendly software tool ASITIC (Analysis and Simulation of Inductors and Transformers for Integrated Circuits). This tool allows circuit and process engineers to design and optimize the geometry of on-chip inductive devices and the IC process parameters affecting their electrical characteristics.

ROBERT G. MEYER

Acknowledgments

The authors wish to acknowledge the financial support of this work by the U.S. Army Research Office (ARO) and the Defense Advanced Research Projects Agency (DARPA).

This book is dedicated to my parents, Afsar and Mohammad.

I
ANALYSIS AND SIMULATION OF PASSIVE DEVICES

Chapter 1

INTRODUCTION

1. INTRODUCTION

The wireless communication revolution has spawned a revival of interest in the design and optimization of radio transceivers. Passive elements such as inductors, capacitors, and transformers play a critical part in today's transceivers. In this book we will focus on the analysis and applications of such devices. In particular, we will demonstrate techniques for calculating the loss when such devices are fabricated in the vicinity of conductive substrates such as Silicon.

2. PASSIVE DEVICES IN EARLY INTEGRATED CIRCUITS

Until recently, passive devices, especially in integrated form, played a relatively minor role in Si integrated circuits in comparison with active devices such as transistors. The most important reason for this can be attributed to the size difference. Active devices were continually shrinking and occupying less and less chip area whereas passive devices remained large. At low frequencies, circuit designers employed simulated passive devices as much as possible to make their products more compact and reliable.

While it was possible to fabricate small values of capacitance on-chip, inductors were virtually impossible due to the large physical area required to obtain sufficient inductance at a given frequency. This was compounded by the losses in the substrate which made it virtually impossible to fabricate high quality devices. Small Si die were desirable to keep costs low and to improve reliability since larger die resulted in lower yields [Gray and Meyer, 1993].

When passive devices were needed, usually they were connected externally on-board rather than on-chip. This is possible as long as few external components are needed and the package parasitics are negligible in comparison with

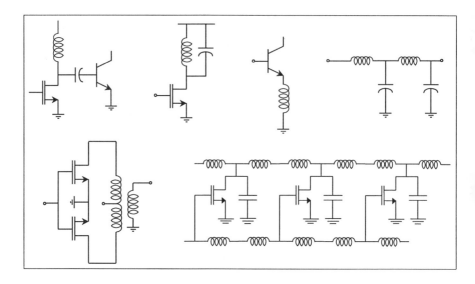

Figure 1.1. Applications of passive devices in Si IC building blocks. (a) Impedance matching. (b) Tuned load. (c) Emitter degeneration. (d) Filtering. (e) Balun. (f) Distributed amplifier.

the external electrical characteristics of the device. Take, for instance, a VCO at 100 MHz versus a VCO at 10 GHz. At 100 MHz a typical tank inductance value will be on the order of 100 nH whereas at 10 GHz the tank inductance is around 1 nH. To access a 1 nH inductor externally is impossible in standard low cost packaging since the package pin and bond wire inductance can exceed 1 nH. Also, as more and more functionality is integrated on-chip, more and more passive devices are needed requiring larger packages which increase the cost. These issues have led to a surge of recent interest in integrated passive devices.

3. APPLICATIONS OF PASSIVE DEVICES

Passive devices, such as inductors and transformers, play an integral part in the performance of circuit building blocks, especially at high frequency. Inductors can be avoided at lower frequencies by using simulated inductances employing active devices. Simulated inductors are more difficult to realize at higher frequencies as active device gain drops. In addition, simulated inductors have finite dynamic range, require voltage headroom to operate, and inject additional noise into the circuits. These limitations place a severe restriction on their application, especially in highly sensitive analog building blocks.

In Fig. 1.1 we see several common applications of inductors, capacitors, and transformers in wireless building block circuits. In (a) we see a narrow-band impedance matching example. Here the input impedance of the second transistor is matched to an optimal impedance value desired by the driving transistor. For instance, in a power amplifier the input impedance of a large output stage device is low due to the capacitance and to obtain sufficient power gain this low impedance is transformed into a larger value. Impedance matching allows circuit designers to obtain minimal noise, maximum gain, minimal reflections, and optimal efficiency when designing circuit building blocks such as low-noise amplifiers (LNAs), frequency-translation circuits (mixers), and amplifiers.

In (b) we see an LC tuned load. A tuned load can take the place of a resistive load to obtain gain at high frequency. The advantages are clear as an LC passive is less noisy than a resistor, consumes less voltage headroom, and obtains a larger impedance at high frequency. A resistive load is always limited by the RC time constant which limits the frequency response. Tuned loads are also a critical component of oscillators. The LC tank tunes the center frequency of the oscillator and the intrinsic Q allows the tank to oscillate with minimal power injection (and hence noise) from the driving transistor.

In (c) an inductor is used as a series-feedback element. Series feedback can be used to increase the input impedance, stabilize the gain, and lower the non-linearity of the amplifier. By using an inductor in place of a resistor, less voltage headroom is consumed, and less additional noise is injected into the circuit. The inductance can also be used to obtain a real input impedance at a particular frequency, thus providing an impedance match at the input of the amplifier.

In (d) inductors and capacitors are used to realize a low-pass filter. Filters of this type are superior to active filter realizations such as gm-C or MOSFET-C filters as they operate at higher frequencies, have higher dynamic range due to the intrinsic linearity of the passive devices, and inject less noise while requiring no DC power to operate.

In (e) we see a center-tapped transformer serving as a balun, a device which converts a differential signal into a single-ended signal to drive external components. Differential operation is advantageous in the on-chip environment due to the intrinsic noise rejection and isolation. Off-chip components, such as SAW filters, though, are single-ended and a balun is needed to convert external single-ended signals to on-chip differential signals.

Finally, in (f) we see inductors and capacitors forming an artificial transmission line in a distributed (traveling-wave) amplifier. Since the LC network acts like a transmission line, it has a broadband response. A wave propagating on the gate-line is amplified and transferred onto the drain line. If the wave speed on the drain line matches the gate line, the signals on the drain line add in phase and the drain line delivers power into a matched load.

6 INDUCTORS AND TRANSFORMERS FOR SI RF ICS

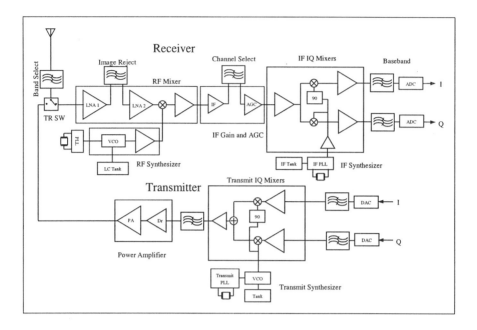

Figure 1.2.   Traditional superheterodyne transceiver architecture.

4.   WIRELESS COMMUNICATION

The wireless transceiver serves as an excellent example of a system which employs passive devices. In recent times, several factors have contributed to the possibility of portable wireless communications. Continuing technology improvements have enabled low-cost Si circuits to operate in the 1–10 GHz frequency range. In this frequency range efficient portable antennas can be realized since the free-space wavelength is on the order of centimeters. Higher frequencies also allow higher bandwidths to be realized for increased throughput or an increase in the number of users sharing the spectrum. Furthermore, by limiting the transmit power, transceivers which are physically remote can reuse the same spectrum with minimal interference leading to the cellular concept of communication. Finally, at these higher frequencies the critical passive elements, such as inductors and capacitors, are small enough to be realized on-chip.

To see the importance of passive devices, consider the simplified block diagram of a traditional superheterodyne transceiver shown in Fig. 1.2. Note that this transceiver is realized as several different chips or modules and many components of this transceiver are off-chip discrete components. This transceiver is thus bulky and expensive. The goal of many research projects has been to realize this transceiver in a more integrated form [Rudell et al., 1997, Rofougaran

Since the vertical base width of a bipolar transistor W_B is determined by a diffusion process whereas the lateral channel length L of an MOS transistor is determined by lithographic processes, bipolar transistors have enjoyed a superiority in speed. New advances in lithographic technology, though, have narrowed the channel length tremendously as MOS technology is the core of the worldwide digital market. Even taking into account that narrow channel device f_T have $1/L$ dependence as opposed to the $1/L^2$ dependence, CMOS technology is today a viable and cost-effective alternative to both bipolar and GaAs and will probably continue to be so for the next decade or longer.

From the perspective of cost, CMOS is the clear winner. But from the perspective of integrated passive devices, GaAs is clearly superior to standard Si. The reason for this stems from the insulating nature of the GaAs substrate which allows very high Q factor passives to be realized on-chip. The dominant limitation in performance is the metal conductivity whereas in Si the dominant loss mechanism at high frequency is the conductive substrate. Electromagnetic energy couples to the substrate and the lossy nature of the Si substrate limits the Q severely.

This is especially the case when the substrate is heavily conductive, as with epi CMOS substrates. Here, magnetically induced eddy currents in the substrate can be a dominant loss mechanism. For moderately conductive substrates, though, electrically induced substrate currents are the dominant loss mechanisms at frequencies below the self-resonant frequency of the device.

6. CONTRIBUTIONS OF THIS RESEARCH

This research has focused on the analysis, design, and applications of passive devices. In Part I emphasis is placed on analyzing inductors and transformers. The solution techniques developed can be easily applied to integrated capacitors and resistors. A general approach is developed from Maxwell's equations to determine the partial inductance and capacitance matrix of an arbitrary arrangement of conductors situated on top of a stratified conductive substrate. The lossy nature of the underlying substrate and losses in the conductors leads to complex capacitance and inductance matrices. These techniques have culminated in a design and analysis software tool called *ASITIC*, "Analysis and Simulation of Inductors and Transformers for ICs." These general techniques can also be applied to extracting the substrate coupling occurring between metal structures residing in the substrate or in the oxide layers on top of the Silicon. This technique is a direct extension of [Gharpurey and Meyer, 1996].

In Part II of the book, we focus on some key applications of passive devices, such as voltage-controlled oscillators and power amplifiers. We show that key performance parameters such as phase noise and power amplifier efficiency depend on the quality of the passive devices. Thus, it is critical to be able to predict the performance of such structures.

Chapter 2

PROBLEM DESCRIPTION

1. DEFINITION OF PASSIVE DEVICES

Consider an arbitrary black box shown in Fig. 2.1 with an arbitrary number of externally accessible terminals. Sample "black" boxes are also shown in the figure. We would like to categorize each black box as passive or active.

Intuitively, the distinction between passive and active devices is clear. While passive devices can only consume or store energy, active devices can also supply energy. Thus, with active devices one can obtain power gain though this is impossible for a passive device. In other words, if we inject power into any terminal of a passive device and observe the power flowing into an arbitrary impedance at another terminal of the device, the real power is necessarily less than or equal to the power injected. If we lift this restriction, the device is active.

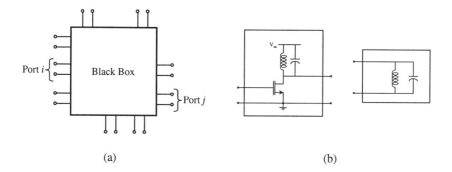

Figure 2.1. (a) An arbitrary black box with externally accessible ports. (b) The contents of two example "black boxes".

Typical examples of passive devices include linear time-invariant resistors of finite resistance $R > 0$ which dissipate energy, ideal time-invariant linear capacitors and inductors, which only store energy. While the above elements are linear elements, passive elements such as diodes can be non-linear[1]. Even though passive devices can have voltage or current gain, they cannot have power gain.

The archetypical example of an active element is of course the transistor and its predecessor, the vacuum tube. Unlike passive devices, such a device can be used to obtain power gain. These devices can be linear or non-linear, but most real world examples are non-linear. The distinction between active and passive is more difficult to make when time-varying elements are employed. For instance, a simple parametric amplifier constructed from a sinusoidally time-varying capacitor in shunt with an RLC tank forms an active device.

A precise definition of passive elements is given in [Desoer and Kuh, 1969]. Given a one-port with port voltage $v(t)$ and port current $i(t)$, the one-port is said to be passive if

$$\int_{t_0}^{t} v(t')i(t')dt' + \mathcal{E}(t_0) \geq 0 \tag{2.1}$$

where $\mathcal{E}(t_0)$ is the energy stored by the one-port at time t_0. With application of this definition, one can clearly delineate an element as passive or active.

At high frequencies it is usually more convenient to discuss the scattering matrix S [Carlin and Giordano, 1964]. The power dissipated by an arbitrary port of an n-port is given by

$$P_k = P_{ik} - P_{rk} \tag{2.2}$$

where the subscript i denotes incident power and r denotes reflected power. In terms of the incident and reflected waves we have

$$P_k = a_k \bar{a}_k - b_k \bar{b}_k \tag{2.3}$$

summing over all ports we obtain the total power

$$P = \sum_{k=1}^{n} P_k = \bar{a}^T a - \bar{b}^T b \tag{2.4}$$

but by the definition of the S matrix $b = Sa$ so that

$$P = \bar{a}^T a - \bar{a}^T \bar{S}^T Sa = \bar{a}^T (I - \bar{S}^T S) a = \bar{a}^T Q a \tag{2.5}$$

where Q is the dissipation matrix. Note that

$$\bar{Q}^T = \bar{I}^T - \overline{(\bar{S}^T S)}^T = I - \bar{S}^T S = Q \tag{2.6}$$

and so Q is a Hermitian matrix. By definition of passivity, we have

$$P = \bar{a}^T Q a \geq 0 \qquad (2.7)$$

Clearly, then, if the matrix Q is positive definite or positive semi-definite, the matrix S corresponds to a passive network. One can also show that this condition implies [Carlin and Giordano, 1964]

$$0 \leq |s_{ij}| \leq 1 \qquad (2.8)$$

This can be deduced intuitively in the following manner. Observe that each diagonal entry of the S matrix is the reflection coefficient when all other ports are matched and each off-diagonal component is the transmission coefficient under matched conditions. For a passive network, the conservation of energy implies that the power reflected from any port must be less than the power injected, thus $|\rho|^2 \leq 1$ and the power transmitted similarly must satisfy the same condition, $|\tau|^2 \leq 1$. This implies all matrix elements have magnitude less than or equal to unity and thus reside in the unit circle in the complex plane [2]. An active device consists of a black box with sources, and thus the matrix elements may have magnitude greater than unity. It follows that the real part of the input port impedances may be negative, something that may never occur for a passive network.

1.1 STABILITY AND PASSIVITY

Passivity is closely related to stability [Desoer and Kuh, 1969]. Again, this is clear intuitively as any passive device must be stable by the conservation of energy. In other words, one can never construct an unstable device with purely passive devices. Conditional stability, as with an oscillating LC tank, is possible as long as the initial conditions supply some energy to the tank. It can be shown that a passive circuit is indeed stable and this further implies that any natural frequency of a passive network must lie in the closed left-hand plane, and any $j\omega$-axis natural frequency must be simple [Desoer and Kuh, 1969].

1.2 RECIPROCITY

Consider the n-port parameters of the black box. If the volume of the black box of Fig. 2.1 is isotropic and encloses no sources, then by the Lorenz reciprocity theorem of electromagnetics it can be shown [Pozar, 1997] that the impedance matrix Z is symmetric. Such a network is called a reciprocal network. A reciprocal network is also described by a symmetric S matrix since in such a case one can show that

$$S = (\frac{Z}{Z_0} + I)^{-1}(\frac{Z}{Z_0} - I) = (\frac{Z}{Z_0} - I)(\frac{Z}{Z_0} + I)^{-1} \qquad (2.9)$$

where Z_0 is the system impedance, usually 50 Ω, and I is the identity matrix. If Z is symmetric, then the symmetry of S follows.

It should be noted that reciprocity has nothing to do with passivity. While many passive networks are indeed reciprocal, the connection is not obvious. Any linear time-invariant RLCM network is reciprocal. In fact, some of the elements may be active and reciprocity is still satisfied. A gyrator, a passive device, in non-reciprocal. The reciprocity theorem in circuits follows from excluding any network with gyrators, dependent and independent sources.

1.3   THE QUALITY OF PASSIVE DEVICES

In general, the complex power delivered to a one-port black box network at some frequency ω is given by [Pozar, 1997]

$$P = \frac{1}{2} \oint_S \mathbf{E} \times \mathbf{H}^* \cdot d\mathbf{s} = P_l + 2j\omega(W_m - W_e) \qquad (2.10)$$

where P_l represents the average power dissipated by the network and W_m and W_e represent the time average of the stored magnetic and electric energy, respectively. One can define the input impedance as follows [Pozar, 1997]

$$Z_{in} = R + jX = \frac{V}{I} = \frac{VI^*}{|I|^2} = \frac{P}{\frac{1}{2}|I|^2} = \frac{P_l + 2j\omega(W_m - W_e)}{\frac{1}{2}|I|^2} \qquad (2.11)$$

If $W_m > W_e$ the device acts inductively whereas if the opposite is true the device acts capacitively.

An important parameter to consider when discussing passive devices is the quality factor. The quality factor has the following general definition

$$Q = 2\pi \frac{E_{\text{store}}}{E_{\text{diss}}} \qquad (2.12)$$

where E_{store} is the energy stored per cycle whereas E_{diss} is the energy dissipated per cycle. Implicit in the above definition is that the device is excited sinusoidally. From (2.11) with T equal to the cycle time

$$Q = 2\pi \frac{(W_m + W_e)}{P_l \times T} = \frac{\omega(W_m + W_e)}{P_l} \qquad (2.13)$$

The higher the Q factor, the lower the loss of a passive device. This definition is most pertinent when discussing inductors or capacitors as such devices are meant to store energy while dissipating little to no energy in the process. Thus ideal inductors and capacitors have infinite Q whereas practical devices have finite Q. Applying the above definition to an ideal inductor L where $W_e \equiv 0$ in series with a resistor R, one obtains $Q = \omega L/R$ and similarly to an ideal capacitor C where $W_m \equiv 0$ in series with a resistor, $Q = (\omega CR)^{-1}$.

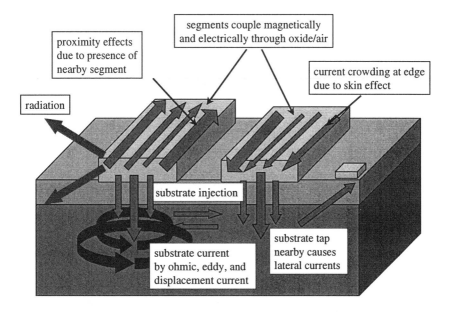

Figure 2.2. Various loss mechanisms present in an IC process.

Physically, the lossy nature of passive devices is rooted in physical phenomena which convert electrical energy into other, unrecoverable forms of energy. Processes which increase entropy are not reversible. For instance, a resistor converts electrical energy into heat. A light bulb converts electrical energy into light and heat. An antenna also converts electrical energy into radiating electromagnetic energy. One can therefore distinguish between passive devices which increase entropy while conserving energy, like a resistor or an incoherent light source, and other devices which conserve energy but do not increase entropy, such as an ideal laser. An ideal laser converts electrical energy into a coherent emission of monochromatic photons. In reality, any physical laser will emit photons with a Lorentzian distribution of energies and thus the entropy of the system increases.

2. LOSS MECHANISMS

The Q factor of integrated passive devices is largely a function of the material properties used to construct the ICs. Specifically, the semiconductor substrate and metal layers used to build the device play the most important roles. The various loss mechanism are summarized in Fig. 2.2 and discussed further below.

16 INDUCTORS AND TRANSFORMERS FOR SI RF ICS

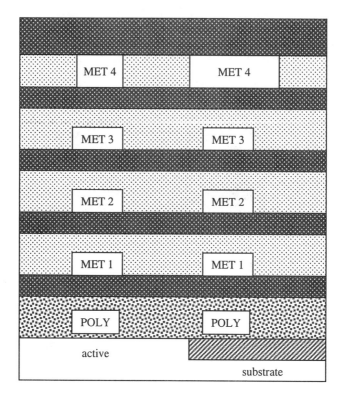

Figure 2.3.   Cross-section of metal and polysilicon layers in a typical IC process.

2.1 METAL LOSSES

Passive devices such as inductors and capacitors are constructed from layers of metal, typically aluminum, and polysilicon layers. Hence, the conductivity of such layers plays an integral part in determining the Q factor of such devices, especially at lower frequencies. For instance, a capacitor is constructed by placing two metal conductors in close proximity. Reactive energy is stored in the electric field formed by the charges on such conductors. Since the metal layers are not infinitely conductive, energy is lost to heat in the volume of the conductors. This loss can be represented by a resistor placed in series with the capacitor. Similarly, an inductor is wound using metal conductors of finite conductivity. Most of the reactive energy is stored in the magnetic field of the device, but energy is also lost to heat in the volume of the conductors.

Fig. 2.3 shows a cross-section of the metal layers of a typical modern IC process. Most processes come with three or more interconnection metal layers. This may include one or two layers of polysilicon as well. Some modern CMOS processes include up to eight metal layers, with the top layer separated from the substrate by ~ 10 μm of oxide.

Most IC metal layers are constructed from aluminum which has a room temperature conductivity of $\sigma = 3.65 \times 10^7$ S/m. Typical metal layers have a thickness ranging from .5 μm to 4 μm, resulting in sheet resistance values from 55 $m\Omega/\square$ to 7 $m\Omega/\square$. Even though silver, copper, and gold are more conductive, at $\sigma_{Ag} = 6.21 \times 10^7$ S/m, $\sigma_{Cu} = 5.88 \times 10^7$ S/m, and $\sigma_{Au} = 4.55 \times 10^7$ S/m, aluminum is the more compatible metal in the IC process. Even though Aluminum is prone to spiking and junction penetration [Jaeger, 1993], it is usually mixed with other metals such platinum, palladium, titanium, and tungsten to overcome these limitations. Electromigration in Al is another problem, setting an upper bound on the maximum safe current density. Although electromigration with AC currents is less problematic, it remains one of the important limitations preventing integration of "high-power" passives on Si, such as the matching networks at the output of a power amplifier. The necessary metal width would require excessively large areas resulting in low self-resonant frequencies.

Many IC processes geared for wireless communication applications are now providing a thick top-metal layer option for constructing inductors. Such a metal layer is also useful for high-speed digital building blocks and clock lines and thus this option is widely available in digital processes as well. This top metal layer may also reside on top of an extra thick insulator for minimum capacitance [Kamogawa et al., 1999]. On the other hand, the wealth of interconnection opens up the possibility of designing structures with many different metal layers. This has the added benefit of requiring no extra processing steps as such "3D" interconnection is available with most modern CMOS processes.

At increasingly higher frequencies, even in the absence of the substrate, the current distribution in the metal layers changes due to eddy currents in the metallization, also known as skin and proximity effects, current constriction, and current crowding. At any given frequency, alternating currents take the path of least impedance. Currents tend to accumulate at the outer layer or skin of conductors since magnetic fields of the device penetrate the conductors and produce opposing electric fields within the volume of conductors. When the effective cross-sectional area of the conductors decreases at increasing frequencies, the current density increases, converting more energy into heat. For an isolated conductor, the magnetic fields originate from the conductor itself (the self-inductance). This increase in AC resistance is know as skin effect and typically follows a $\sqrt{f}$ functional dependence. This rate of increase can be traced to the effective depth of penetration δ of the current since the effective area is a function of the skin depth[3]

$$\delta = \sqrt{\frac{2}{\omega\mu\sigma}} \qquad (2.14)$$

In a multi-conductor system, the magnetic field in the vicinity of a particular conductor can be written as the sum of two terms, the self-magnetic field and the neighbor-magnetic field[4]. Thus, the increase in resistance of any particular conductor can be attributed not only to skin effect but also to proximity effects, the effect of nearby conductors. If nearby conductors enhance the magnetic field near a given conductor, the AC resistance will increase even further and this is the case for a spiral inductor. On the other hand, if the nearby fields oppose the field of a given conductor, as is the case in a transformer, the AC resistance will decrease as a result.

2.2 SUBSTRATE INDUCED LOSSES

Integrated passive devices must reside near a conductive Si substrate. The substrate is a major source of loss and frequency limitation and this is a direct consequence of the conductive nature of Si as opposed to the insulating nature of GaAs. The Si substrate resistivity varies from 10 $k\Omega$-cm for lightly doped Si (10^{13} atoms/cm^3) to .001 Ω-cm for heavily doped Si (10^{20} atoms/cm^3). In fact, to combat these substrate induced losses, some researchers propose removing the substrate from under the device by selective etching [Chang et al., 1993] [Lopez-Villegas et al., 2000].

The conducting nature of the Si substrate leads to various forms of loss, namely conversion of electromagnetic energy into heat in the volume of the substrate. To gain physical insight into the problem, we can delineate between three separate loss mechanisms. First, electric energy is coupled to the substrate through displacement current. This displacement current flows through the substrate to nearby grounds, either at the surface of the substrate or at the back-plane of the substrate. Second, induced currents flow in the substrate due to the time-varying magnetic fields penetrating the substrate. These magnetic fields produce time-varying solenoidal electric fields which induce substrate currents. These currents are show in Fig. 2.4 for the case of a spiral inductor. Note that electrically induced currents flow vertically or laterally, but perpendicular to the spiral segments. Eddy currents, though, flow parallel to the device segments[5].

Finally, all other loss mechanisms can be lumped into radiation. Electromagnetically induced losses occur at much higher frequencies where the physical dimensions of the device approach the wavelength at the frequency of propagation in the medium of interest. This frequency is actually difficult to quantify due to the various propagation mechanisms of the substrate. For instance, if we consider propagation into air, the the free-space wavelength is the appropriate

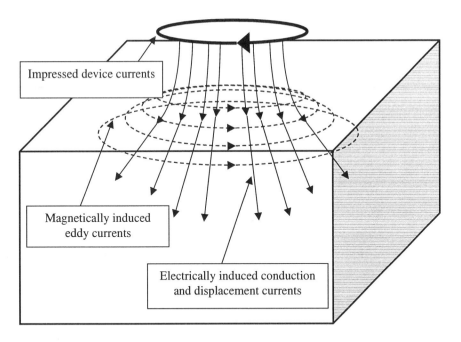

Figure 2.4. Schematic representation of substrate currents. Eddy currents are represented by the dashed lines and electrically induced currents by the solid lines.

factor. Even at 10 GHz, the wavelength in air is 3 cm, much larger than any RF inductor or capacitor. Even at 100 GHz, the wavelength is now 3 mm, still much larger than any device at this frequency. Thus, we can safely ignore the electromagnetic propagation into the air.

Efficient electromagnetic propagation into the substrate, though, occurs at lower frequencies due to the lower propagation speed, roughly at a factor of $\sqrt{\epsilon_{Si}}$ lower due to the diamagnetic nature of Si. Since $\epsilon \approx 11.9$ in Si, this is slightly slower from propagation in air. Furthermore, due to the lossy nature of the substrate, waves traveling vertically into the surface of the substrate are heavily attenuated. Waves traveling along the surface of the substrate, though, can propagate partially in the lossless oxide and partially in the substrate. For a lightly doped substrate, the wave propagation behaves like a "quasi-TEM" mode. As the substrate is made heavily conductive, the wave is constrained to the oxide and the substrate acts like a lossy ground plane. This is the so-called "skin effect" mode of propagation [Hasegawa et al., 1971]. There is a third kind of possible excitation, the "slow-wave" mode of propagation, where the effective speed of propagation is orders of magnitude slower than propagation in free space [Hasegawa et al., 1971].

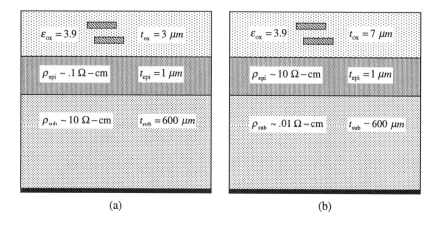

Figure 2.5. Cross-section of typical (a) bipolar and (b) CMOS substrate layers.

2.2.1 SI IC PROCESS SUBSTRATE PROFILE

In Fig. 2.5 typical bipolar and CMOS process substrate profiles are shown. Each substrate consists of one or more layers of Si or a compatible material[6]. Layers of varying conductivity are added to the bulk substrate by various g processes, such as diffusion, chemical vapor deposition and growth, epitaxy, and ion implantation. Various layers of oxide (SiO_2 for instance) and polyimide are also grown to provide insulation from the substrate and between metal layers.

In general, the more conductive the substrate layers, the more detrimental the resulting losses. It is therefore no surprise that intrinsic Si substrates [7] result in the lowest losses [Park et al., 1997b]. Due to the close proximity of the Si substrate to the inductors and transformers residing in the metal layers, the case of an infinitely conductive substrate is also problematic. For a heavily conductive substrate, the magnetic and electric fields do not penetrate the substrate appreciably and even though no substrate induced losses occur, the surface currents flowing in the substrate, acting like "ground-plane" currents, produce opposing magnetic fields which tend to drive the inductance value of coils to low non-usable values.

Therefore, given the choice, designers of ICs and process engineers should ensure that as few as possible conductive substrate layers appear under or near an inductor. This is unfortunately not always possible due to planarization constraints. Furthermore, the thickest possible oxide should be realized under the device to minimize the substrate capacitance. This not only minimizes the losses, but also maximizes the self-resonant frequency of the device. In the limit, self-resonance will occur due to interwinding capacitance as opposed to substrate capacitance. Since interwinding capacitance can be controlled by increasing the metal spacing, this gives the IC designer more control over the passive device behavior.

Most bipolar and BiCMOS substrates come with a standard 10–20 Ω-cm substrate. With this value of resistivity, electrically induced losses dominate the substrate losses in the 1–10 GHz frequency range [Ruehli, 1972]. This is also the case for bulk CMOS substrates with the same range of resistivity. In such a case, one must ensure that no conductive n- or p-wells appear below the device. This may require a special mask to block the dopants in the well creation process, especially for a twin-well CMOS process. To minimize the chance of latch-up, many modern CMOS processes begin with a heavily conductive thick substrate about 700 μm thick and grow a thin epitaxial layer of resistive Si on the surface to house the wells. This is unfortunate for RF/microwave circuits as the bulk substrate can be as conductive as 10^4 S/m and this can be a major source of substrate induced losses due to eddy currents.

The back-plane of the substrate may or may not be grounded. Even if it is physically grounded for DC signals, AC signals are constrained to flow within several skin depths δ and this factor is a strong function of the conductivity. For heavily conductive substrates, currents are constrained to flow at the surface of the substrate at high frequencies whereas for moderately conductive substrates currents flows deep into the substrate and into the back-plane ground.

A physical ground may be realized if the die (chip) resides in a package with a conductive ground plane, or if the die is bonded directly onto a board, and a conductive epoxy cement glue is used. Although the epoxy is not conductive, metals can be mixed in to produce a conductive solution. In the modeling of the substrate this can be an important factor in enforcing the boundary conditions surrounding the chip.

Unless the substrate thickness is reduced substantially, the conductive backplane ground is sufficiently distant not to appreciably influence the electromagnetic behavior of the inductor. On the other hand, some packages use "down-bonds," bond wires from the Si die to the package grounded "paddle". To minimize the bond wire length, the substrate thickness is reduced in postprocessing steps. This has further benefits for a packaged power amplifier since a thinner substrate also has better thermal conductivity. For such a thin, moderately conductive grounded substrate, one must take into account ground "image" currents which can reduce the inductance value and serve as a further loss mechanism [Krafcsik and Dawson, 1986]. If the substrate is sufficiently conductive such that the skin depth δ is much less than the substrate thickness, then image currents will be confined to flow at the substrate surface.

3. DEVICE LAYOUT

In this section we will discuss various ways to lay out inductors using the planar metallization layers of a typical IC process. Off-chip inductors are usu-

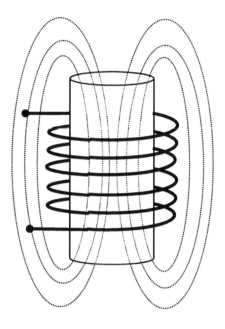

Figure 2.6. The typical coil inductor.

ally realized as a solenoidal coil or toroid, as shown in Fig. 2.6. Each additional turn adds to the magnetic field in phase with the previous turn. The magnetic energy is stored mostly in the inner core of each winding. The inductance is largely a function of the area of the loop and the number of turns in the winding typically resulting in an N^2 dependence.

3.1 PLANAR INDUCTOR STRUCTURES

Since on-chip inductors are constrained to be planar, the typical solution is to form a spiral, as shown in Fig. 2.7. Since some IC processes constrain all angles to be 90°, a square version of the spiral, shown in Fig. 2.8, is a popular alternative. A polygon spiral, as shown in Fig. 2.9, is a compromise between a purely circular spiral and a square spiral.

In designing integrated circuits it is sometimes convenient to tap an inductor at some arbitrary point. While this is certainly possible, as more than one metal layer is present, it is sometimes necessary to tap a spiral in the center, especially for differential circuits. In a spiral it is difficult to find such a symmetric center point since the electric fields on the outer turns tend to fringe and thus the "inductive" center does not correspond to the "capacitive" center. Also, the "inductive" center does not correspond to the "resistive" center due to the non-uniform mutual magnetic coupling. To solve this problem, some researchers

Figure 2.7. A circular spiral inductor.

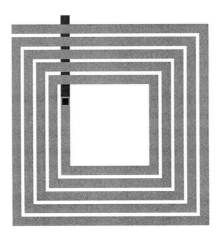

Figure 2.8. A square spiral inductor.

have proposed symmetric structures, such as that shown in Fig. 2.10. Note that each turn involves a metal-level interchange, a process that requires vias. A different center-tapped structure proposed by [Kuhn et al., 1995] requires only one metal interchange. This structure is very similar to a inter-digited planar transformer structure shown in Fig. 2.13. These structures have a natural geometric center which coincides with the electrical center point. This is needed in differential circuits as such points can be grounded or connected to supply

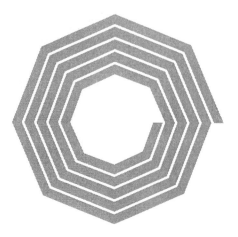

Figure 2.9. A polygon spiral inductor.

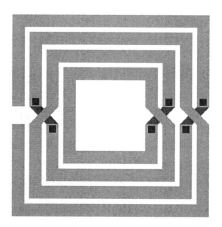

Figure 2.10. A symmetric spiral inductor.

without disturbing the differential signal. Circular or polygon versions are also possible, as shown in Fig. 2.11.

3.2 NON-PLANAR INDUCTOR STRUCTURES

Up to now we have only considered planar structures even though modern IC processes offer many metal layers. Two simple approaches in utilizing the metal layers are to connect multiple spiral inductors in series or in shunt. While N spirals in series increase the series resistance by a factor of approximately N

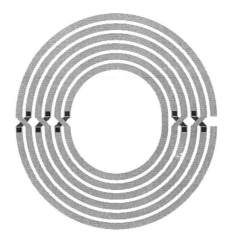

Figure 2.11. A symmetric polygon spiral inductor.

(neglecting via resistance), the inductance value increases faster due to the mutual magnetic coupling. At low frequencies where the current flowing through each series connected spiral I_j is equal, the effective inductance of N series connected coupled inductors is

$$L_{se} = \sum_{i=1}^{N} L_i + 2 \sum_{i=1}^{N} \sum_{j \neq i} M_{ij} \quad (2.15)$$

where M_{ij} is the mutual magnetic coupling between each series connected spiral i and j. Thus, the series connection approaches an N^2 increase in inductance and the Q factor can potentially increase by a factor of N for the case of perfectly coupled spirals ($k = 1$).

Alternatively, in the shunt connection, the series resistance drops by a factor of N (assuming equal resistivity in each metal layer and uniform current distribution among the coils) whereas the inductance of mutually coupled inductors drops to[8]

$$L_{sh} = \frac{1}{\sum_{i=1,j=1}^{N} K_{ij}} \quad (2.16)$$

where the matrix K is the inverse of the partial inductance matrix M. This result certainly agrees with the case of a diagonal matrix M corresponding to zero coupling since in such a case we have the familiar result for parallel inductors

$$L_{sh,k=0} = \frac{1}{\sum_{i=1}^{N} \frac{1}{M_{ii}}} \quad (2.17)$$

For the case of two coupled inductors we have a simpler relation

$$L_{sh} = \frac{1}{2}\frac{L_1 L_2 - M^2}{L_1 + L_2 - 2M} \qquad (2.18)$$

For the case of perfectly coupled equal value inductors, with $k = +1$, the above result yields $L_{sh} = L$. For the general case, the matrix M is singular but by symmetry the current flowing through all inductors is equal so the voltage across the jth inductor gives

$$V_j = \sum_{k=1}^{N} sM_{jk}I_k = \frac{I}{N}\sum_{k=1}^{N} sM_{jk} = \frac{I}{N}sL\sum_{k=1}^{N} 1 = sLI \qquad (2.19)$$

and so we have

$$L_{sh} = L_1 = L_2 = \cdots = L_N \qquad (2.20)$$

In this limit, the Q factor also improves by a factor of N due to the drop in series resistance. In practice, both the series and shunt connection offer a Q improvement close to the theoretical limit due to the tight coupling achievable in the on-chip environment. These benefits, though, only occur at low frequencies where the above assumptions hold.

At high frequencies, both approaches are liable to reduce the Q factor over the case of a single layer coil due to the capacitive and substrate effects. The series connection suffers from high interwinding capacitance which lowers the self-resonance frequency lowering the maximum frequency of operation of the device. The shunt connection moves the devices closer to the substrate where capacitive current injection into the substrate may dominate the loss of the device.

Another approach is to attempt to realize a lateral coil on-chip by using the top and bottom metal layers and vias as the side. This approach has been successfully demonstrated in [Young et al., 1997] using special post-processing steps to realize sufficient cross-sectional area in the coil. This approach has the added advantage of positioning the magnetic fields laterally to the substrate where eddy currents are reduced. In a vertical coil or spiral, the magnetic field is strongest at the center of the coil. Due to the finite conductivity of the substrate, these changing magnetic fields leak into the substrate and produce eddy currents. Eddy currents can be a significant source of loss and this technique might be an effective method to combat this loss mechanism. Standard monolithic integration, though, has failed to produce a coil with sufficient cross-sectional area to produce significant Q. Some innovative strategies around this are to employ a combination of metal and bond wires to realize the coil [Lee et al., 1998b]. The work of [Lee et al., 1998b] claims tight tolerance on the inductance value which is a primary concern of using bond wires alone[Craninckx and Steyaert, 1995].

Problem Description   27

Figure 2.12. A tapered spiral inductor.

3.3 TAPERED SPIRALS

A tapered spiral is shown in Fig. 2.12. The metal pitch and spacing are varied to minimize the current constriction at high frequency [Niknejad, 1997, Lopez-Villegas et al., 2000]. Current constriction, or skin effect, is non-uniform as a function of the location in the spiral due to the non-uniformity of the magnetic field. At low frequencies the current is nearly uniform whereas at high frequency non-uniform current flows due to proximity effects.

The magnetic field is strongest in the center of the spiral [Craninckx and Steyaert, 1997] and thus the time-varying magnetic field produces eddy currents of greatest strength in the volume of conductors near the center of the device. Since at high frequency current constriction limits the current to the outer edges of the conductors, conductor width does not have as strong an influence on minimizing metal losses as at low frequency. For this reason [Craninckx and Steyaert, 1997] advocate removing the inner turns to produce a "hollow" spiral. Another approach is to decrease the width of the inner turns and to effectively move these turns closer to the outer edge. This approach contrasts with the approach suggested by [Lopez-Villegas et al., 2000] where the sum of the metal pitch and spacing, $W + S$, is kept constant.

Tapering is most effective when substrate losses are negligible, as in the case of an insulating substrate. This stems from the fact that current constriction dominates the metallization losses at high frequencies when the substrate losses are also significant. On the other hand, tapering can be an effective means of increasing the self-resonant frequency of a device by decreasing the cross-sectional area of the device.

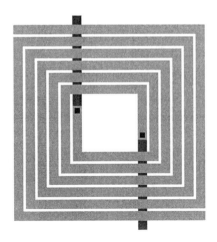

Figure 2.13. A planar square spiral transformer.

3.4 TRANSFORMERS

On-chip transformers are realized very similarly to inductors. To maximize the coupling factor k, two inductors can be interwound as shown in Fig. 2.13. Polygon spirals can be similarly interwound to form transformers. These transformers have equal turns ratio at the primary and secondary. Since the turns ratio n is given by

$$n = k\sqrt{\frac{L_2}{L_1}} \qquad (2.21)$$

one way to realize $n \neq 1$ is to alter the number of turns and metal pitch in the secondary. In addition, to lower the losses in the secondary, turns may be connected in shunt [Long and Copeland, 1997]. Typical practical coupling values in the range of $.6 < k < .8$ can be achieved with planar spirals. Metal-metal transformers utilizing two or more metal layers can save area, but cause asymmetry and increased capacitive coupling between the primary and secondary. The asymmetry can be reduced by utilizing a structure such as the one in Fig. 2.14.

If center-tapped transformers are desired, such as in a balun, a structure such as Fig. 2.15 serves well. This structure is formed by interwinding two center-tapped coils of Fig. 2.10.

3.5 SHIELDED STRUCTURES

In an attempt to shield a device from substrate losses, researchers have proposed building a shield with lower metal layers or polysilicon layers to block

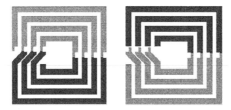

Figure 2.14. An expanded view of a non-planar symmetric spiral transformer. The primary is shown on the left and the secondary on the right. These inductors actually reside on top of one another.

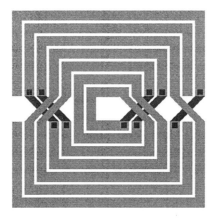

Figure 2.15. A planar symmetric balun transformer.

electromagnetic energy from coupling to the substrate [Yue and Wong, 1997]. While electrostatic shielding works well for capacitors and RF pads, especially at "low-frequencies", the shield must be patterned in the case of inductors so as to avoid or reduce the effects of eddy currents. Due to the close proximity of the device and the shield, using solid metallization would allow "image" eddy currents to flow which would produce an opposing magnetic field. This would reduce the device magnetic energy storage and hence the Q factor. A patterned shield, similar to Fig. 2.16, only allows shield currents to flow perpendicular to the conductive paths of a spiral inductor, thereby preventing the majority of eddy currents which flow parallel to the device[9].

The effects of a shield, or any other "grounding" structure, can be analyzed with the techniques presented in this research. At first glance a patterned shield seems like a very effective means of improving the device Q. This, however,

Figure 2.16. A patterned ground shield.

must be carefully examined for each process and product. There are many obvious problems with a shield, such as drastically reduced self-resonant frequency. There are also more subtle problems with a shield.

To see this, consider the inductor layout as a lossy inductor and a parasitic lossy capacitor, C_s in series with a loss resistor R_s. This frequency-independent representation is certainly valid as the network two-port parameters can be transformed into this equivalent circuit uniquely at a given frequency. In practice, the addition of a few elements, such as a lossless interwinding capacitor, broadbands the model, especially below self-resonance [Niknejad, 1997]. Then the action of the shield is to effectively increase the Q_C factor of the capacitor portion of the equivalent circuit. Consider a series to parallel transformation of this lossy capacitor into C_p in shunt with R_p

$$R_p = (1 + Q_C^2)R_s \qquad (2.22)$$

$$C_p = \frac{Q_C^2}{1 + Q_C^2}C_s \qquad (2.23)$$

Suppose that without a shield the Q_C factor of this capacitor is very low, $Q_C \ll 1$, such that

$$R_{low} \approx R_s \qquad (2.24)$$

$$C_{low} \approx Q_C^2 C_s \qquad (2.25)$$

If R_{low} is then already larger than the parallel equivalent inductor loss resistor, then this capacitor plays a minor role in determining the overall Q and shielding will actually deteriorate the performance of the device. Note that since

$C_{low} \ll C_s$, the effective shunt capacitance is small and does not lower the self-resonant frequency of the device. Clearly, in this case shielding does not help. Now consider the other extreme where the resistor R_{low} actually loads the tank significantly. Then clearly increasing the capacitor Q_C factor helps since for $Q_C \gg 1$

$$R_{high} \approx Q_C^2 R_s \qquad (2.26)$$
$$C_{high} \approx C_s \qquad (2.27)$$

This occurs, on the other hand, at the expense of lowered self-resonance since now the parasitic capacitor loads the tank. So we see the shield may potentially improve the overall tank Q but at the cost of reducing the usable frequency range of the device. Another approach, discussed in [Burghartz et al., 1997], surrounds the device with an open halo of substrate contacts. This has the added benefit of increasing the capacitor Q_C without loading the tank capacitance significantly. Measurements by other researchers have also corroborated these findings [Yoshitomi et al., 9998].

Another potential problem with the shield arises due to finite non-zero ground inductance. Since most packaged ICs suffer both package and bond wire inductance, the actual zero potential resides off-chip and there is considerable "ground bounce" on-chip. A typical IC has several inductors and transformers and if they are shielded, they are all effectively tied to a common non-zero impedance point. Thus a parasitic coupling path exists from device to device.

For instance, in an amplifier, this intra-block leakage can either lower the gain (as negative feedback) or cause instability (as positive feedback). Inter-block coupling, on the other hand, can produce spectral leakage and spurs, jamming and reduced SNR (gain compression), or mode-locking.

Any IC, of course, suffers from parasitic substrate coupling but shielding increases the substrate capacitance value and removes the resistive isolation between the devices.

3.6 VARACTORS (REVERSE-BIASED DIODES)

Varactors, or variable capacitors, are usually constructed as back-biased diodes. Diodes are realized as junctions between p-type and n-type doped regions of Si. Briefly, the presence of an n-type region abutting a p-type region creates a large concentration gradient which leads to diffusion. This diffusion forms the so-called "space-charge" or "depletion" region, a charged volume of space surrounding the junction. The charge of this region is due to the charge of immobile dopant sites. Donor or acceptor atoms are easily ionized at room temperature resulting in free mobile carriers. The diffusion process is balanced by the conduction current resulting from the built-in electric field since the charge buildup from the ionized immobile dopants produces an electric field opposing the diffusion. Thus a natural barrier is formed against charge crossing the junction boundary.

If the diode is forward biased beyond a threshold, carriers obtain sufficient energy to cross the barrier. On the other hand, if the diode is reverse-biased, no DC conduction current flows. In reality, a small current will flow even under reverse-biased conditions. One contribution to these "leakage currents" is due to minority carriers which are created on average within a diffusion length of the junction. These carriers are actually propelled by the built-in electric field and cross the junction. Note that this current is relatively bias-independent[10].

While little to no DC current flows, AC displacement currents can flow since the reverse-biased diode acts very much like a capacitor. The thin space-charge region serves as the "insulator" and the n-type and p-type Si regions act as capacitor plates. Electric energy is stored in the built-in field of the space-charge region. Under an abrupt junction assumption, the depletion region thickness is given by [Neudeck, 1989]

$$d_j = \left[\frac{2K_S\epsilon_0(V_{bi} - V_A)}{q}\frac{(N_A + N_D)}{N_A N_D}\right]^{1/2} \quad (2.28)$$

N_D and N_A are the donor and acceptor volume densities and V_{bi} is the built-in potential given by

$$V_{bi} = \frac{kT}{q}ln\left[\frac{N_D N_A}{n_i^2}\right] \quad (2.29)$$

where complete ionization of the donors and acceptor is assumed. Note that under reverse-biased conditions, the spacing d_j is a function of the applied voltage $V_A < 0$. This effect can be exploited to produce a variable capacitor. The physical origin of the dependence of d_j on V_{rev} is simply due to the fact that increasing the voltage difference increases the electric field which requires more charge which in turn comes from extending the space charge region. If $N_D \gg N_A$ then the fractional increase in d_j from the donor side is small compared to the acceptor sites. The opposite, of course, also applies. Thus, in such a case the series resistance of the diode in reverse bias will be dominated by the acceptor side, or the p-type Si, due to its higher resistivity. Note that if both the donor and acceptor densities are increased in an attempt to lower the series losses, the capacitance gets large. While this is desirable, the breakdown voltage of the reverse-biased junction decreases. For avalanche breakdown we have [Neudeck, 1989]

$$V_{BR} = \left(\frac{\mathcal{E}_C T^2 \epsilon_r}{2q}\right)\left[\frac{N_A + N_D}{N_A N_D}\right] \quad (2.30)$$

The same applies for Zener breakdown since decreasing d_j increases the probability that a carrier will tunnel through the potential barrier. Since the operating voltage of ICs is reducing, a proportional decrease in the breakdown voltage is tolerable allowing lower series resistance and thus higher Q values to be realized.

The equivalent capacitance of a junction with cross-sectional area A is thus given by

$$C_j = \frac{\epsilon_r \epsilon_0 A}{d_j} = \frac{\epsilon_r \epsilon_0}{\left[\frac{2 K_S \epsilon_0 (V_{bi} - V_A)}{q} \frac{(N_A + N_D)}{N_A N_D}\right]^{1/2}} \quad (2.31)$$

this can be rewritten as

$$C_j = \frac{C_{j0}}{\left[1 - \frac{V_A}{V_{bi}}\right]^m} \quad (2.32)$$

where C_{j0} is the junction capacitance for the zero applied bias case, $V_A = 0$. The factor m is equal to $1/2$ for an abrupt junction but takes on the value of $1/3$ for a linearly graded junction. For real junctions, one finds that $1/3 < m < 1/2$.

A pertinent factor when designing VCOs and filters with variable cutoff is the derivative of (2.32) with respect to the applied voltage V_A

$$K_{CV} = \frac{C_{j0} m}{V_{bi}(1 - \frac{V_A}{V_{bi}})^{m+1}} \quad (2.33)$$

Increasing this factor is important for maximizing the change in capacitance for a change in the reverse-biased voltage. This maximizes the tuning range of an LC tank, for instance, since

$$\frac{d\omega}{dV} = \frac{-K_{CV}}{2 C_j} \omega_0 \quad (2.34)$$

where $\omega_0 = \sqrt{1/LC_j}$.

3.7 MOS CAPACITORS

Another technique to realize a variable capacitor is through an MOS capacitor. The MOS capacitor has four distinct regions of operation: accumulation, depletion, weak inversion, and strong inversion. In accumulation mode, the capacitance is equal to the oxide capacitance

$$C_{acc} = C_{ox} = \frac{\epsilon_r \epsilon_0 A}{t_{ox}} \quad (2.35)$$

where t_{ox} is the gate oxide thickness and A is the capacitor plate area. This assumes that majority carrier distribution equilibrates much faster than the period of the applied signal. We also assume that all the charge resides at the surface of the substrate (delta-function approximation for charge density) and we neglect the fringing fields. In reality, the charge distribution will not peak at the surface but slightly away from the surface due to quantum mechanical considerations.

In depletion mode, majority carriers at the surface of Si are repelled by the applied electric field and the region is thus depleted of mobile carriers. A depletion region forms in a similar manner to the reverse-biased junction diode considered previously. The capacitance from the gate to the substrate, therefore, has two components, an oxide capacitance in series with a depletion capacitance [Pierret, 1990]

$$C_{dep} = \frac{C_{acc}C_{dep}}{C_{acc} + C_{dep}} = \frac{C_{ox}}{1 + \frac{\epsilon_{r,SiO_2} d_j}{\epsilon_{r,Si} t_{ox}}} \quad (2.36)$$

As the channel region begins to invert, minority carrier charge accumulates at the surface and the capacitance once again approaches the C_{ox} value. This assumes that the rate of minority carrier generation is sufficiently fast to follow the time-varying AC signal. On the other hand, if a MOSFET-C structure is used instead, a large source of minority carrier charge is available from the surrounding source and drain regions. Between inversion and depletion, or in weak inversion, the situation is more complicated as an incremental increase in the applied voltage can result in new field lines terminating on both immobile charges (by an increase in the depletion junction depth) and terminating on newly generated minority carriers.

In the weak inversion region, the capacitance changes rapidly from C_{dep} to C_{ox} due to the exponential build-up of charge at the surface. This results in a very large K_{CV} factor. The quality factor of such a capacitor is limited mainly by the gate series resistance and can be shown to be approximately equal to $R_G/3$[11].

3.8 RESISTORS AND CAPACITORS

Resistors can be implemented in several ways but all essentially involve moving current through a length of material with relatively high resistivity. In a standard bipolar process, diffusion resistors are common. The resistance value can be varied by the application of a bias such as in base-pinch and epitaxial pinch resistors [Gray and Meyer, 1993]. In a standard MOS process we also find diffusion resistors, polysilicon resistors, well-resistors, and the MOS device itself as variable resistors. In a modern bipolar, BiCMOS, or MOS process, most of the above options are available.

Capacitors are realized as either MOS capacitors, as poly-poly "metal" capacitors, metal-insulator-metal (MIM) capacitors, and as poly-diffusion capacitors. For RF applications MIM capacitors are preferred since they result in the lowest losses. Even though standard metal interconnection can be used to construct MIM capacitors, this has two negative consequences. First, it will result in a large uncertainty in the capacitance value due to wide oxide thickness variation

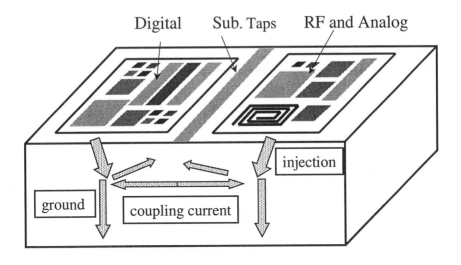

Figure 2.17. Substrate coupling in a mixed-signal system on a chip.

across the wafer. Second, the capacitance value per unit area will be very small due to the thick oxide separating the metal layers. A special MIM capacitor constructed with a well-controlled thin oxide is therefore preferred.

Resistors and capacitors have always been a part of the standard IC process. Thus, great effort has already gone into the analysis and modeling of such structures at lower frequencies. At higher frequencies, though, these structures have many non-ideal effects which must be taken into consideration, such as substrate coupling, self-resonance, and inductance. In this work we will focus on a general interconnection of metal structures above a lossy Si substrate and this generality will allow us to apply our analysis technique to inductors as easily as to capacitors, transformers, and resistors.

4. SUBSTRATE COUPLING

An issue that permeates this research is substrate coupling. Figure 2.17 illustrates the various substrate coupling mechanisms present in an IC environment. Current is injected into the substrate through various mechanisms. Physically large passive devices such as inductors, capacitors, transformers, interconnect and bonding pads inject displacement current in the substrate. This current flows vertically and horizontally to points of low potential in the substrate, such as substrate taps and the back-plane ground. This current couples to other large passive structures in a similar manner.

Active devices also inject current into the substrate, directly and capacitively. Since most active devices are isolated from the substrate by either reverse-biased *pn* junctions or oxide, capacitive substrate current injection and reception occur at high frequencies. Direct current can also be injected into the substrate due to hot electron effects. In a short-channel MOS transistor, for instance, electron-hole pair creation takes place in the high-field pinch-off region near the drain due to collisions. The electric field lines lead one set of the carriers into the substrate. Under high field conditions, a multiplication process (avalanche) can also lead to additional hole-pair creation and result in substantial currents.

Coupling between sensitive analog nodes can lead to instability or gain reduction. Furthermore, in a mixed signal IC the coupling between analog and digital portions of the chip can be very problematic. Digital gates switch in a pseudo-random manner and each gate transition results in some energy transfer to the substrate. This energy can couple to sensitive analog nodes and result in reduced signal-to-noise (SNR) ratios.

Though careful layout techniques and differential operation minimize them, these effects can never be totally eliminated or ignored in the design of mixed-signal ICs. Substrate coupling is treated extensively in [Gharpurey and Meyer, 1996]. In this book we draw from an extension of this work, presented in [Niknejad et al., 1998], to analyze passive device substrate coupling.

Notes

1. This depends on the I-V curve of a diode. A tunnel diode can certainly be biased as to supply energy, becoming an active device.
2. This is a necessary condition for a passive device. However, this condition alone is not sufficient since power gain may occur for a non-matched load or source impedance.
3. This is not a rigorous argument since the skin-depth concept of surface impedance applies strictly to a semi-infinite conductor.
4. This is due to the linearity of Maxwell's Equations.
5. In general the eddy currents are solenoidal whereas the displacement and conductive currents are curl free
6. By this we mean the crystal structures of the adjacent layers are compatible.
7. By intrinsic we mean no intentional dopants are introduced into the substrate and the only conduction occurs through thermionic emission into the conduction band.
8. This result will be established in Chapter 5.
9. Current distributions with zero curl lead to zero magnetic fields.
10. Consider the current associated with a stream of suicidal lemmings. Given that the the cliff is sufficiently high, the suicide rate of lemmings is independent of the height of the cliff.
11. See Appendix A.

Chapter 3

PREVIOUS WORK

1. EARLY WORK

The calculation of inductance has a long history and dates back to early researchers. In fact, the geometric mean distance approximation (discussed in Chapter 5) dates back to Maxwell.

More recently, Ruehli's seminal paper [Ruehli, 1972] on inductance calculation clarified the partial inductance approach and suggested how to apply this technique to the complex integrated circuit environment. Two years later, Greenhouse [Greenhouse, 1974] wrote an important and oft cited paper on the analysis of printed spiral inductors. Greenhouse's approach differed greatly from other researchers as he abandoned the search for an approximate closed-form expression for the inductance of a spiral and opted instead for an expression based on the partial-inductance concept, more appropriate for numerical calculation. Greenhouse's work drew a great deal from the work of Grover [Grover, 1946] and his exhaustive compilation of tables and formulae for calculating the static inductance value of practical configurations, such as filaments, loops, coils, toroids, and spirals.

Other researchers [Weeks et al., 1979] extended the work of Ruehli to calculate the high frequency inductance and skin effect in practical conductors. Ruehli also published a series of papers [Ruehli, 1974] [Ruehli and Heeb, 1992] [Ruehli et al., 1995] where he developed the concept of Partial Element Equivalent Circuits (PEEC), a technique for solving Maxwell's equations which is used extensively throughout this work. The PEEC formulation has also been the core of many other numerical techniques to calculate capacitance [Nabors and White, 1991] and inductance [Kamon et al., 1994a] efficiently.

2. PASSIVE DEVICES ON THE GAAS SUBSTRATE

Several researchers have focused attention on modeling passive devices on the GaAs substrate. Spiral inductors and transformers were especially problematic and much work was done to extend the work of Greenhouse to include distributed effects, losses, and self-resonance.

The research of Cahans [Cahana, 1983] and [Shepherd, 1986] treated the spiral inductor as an interconnection of coupled transmission lines, capturing the distributed nature from the outset. This fundamental approach has been extended by many researchers [Boulouard and Le Rouzic, 1989] [Schmuckle, 1993].

The work of Krafcsik and Dawson [Krafcsik and Dawson, 1986] took a different approach. Instead of treating the spiral segments as distributed elements, lumped elements were used similar to the Greenhouse approach. The distributed nature of the spiral is also included through coupling capacitors excited by phase-delay within the spiral. The reduction in inductance due to ground eddy currents was modeled by image currents. The work of Pettenpaul and colleagues [Pettenpaul and et al., 1988] is along the same lines. The approach models each spiral segment individually, similar to the PEEC approach, and then combines the individual models to a macro two-port representation. These techniques form the foundation for the work presented in this research.

3. PASSIVE DEVICES ON THE SI SUBSTRATE

In [Nguyen and Meyer, 1990] Nguyen and Meyer demonstrated the feasibility of integrating spiral inductors in the Si IC environment. This work was followed by a plethora of experimental and theoretical research aimed at improving the quality factor and integration of on-chip spiral inductors and transformers. This mass research effort can be attributed to the important role of integrated passive elements in realizing a fully-integrated radio transceiver [Gray and Meyer, 1995].

3.1 EXPERIMENTAL RESEARCH

In order to reduce the metal and substrate induced losses, many research efforts have been aimed at modifying the device structure and/or the IC process to minimize losses. Using thick and more conductive metallization minimizes the low frequency losses whereas using a heavily resistive substrate along with a thick oxide layer minimizes the substrate losses. Naturally, using an oxide with a lower dielectric constant or eliminating the substrate entirely helps a great deal, as well.

For instance, to minimize the effects of the substrate, Chang and Abidi [Chang et al., 1993] demonstrated the feasibility of higher inductance and Q factors by removing the Si substrate through selective etching, realizing a 100 nH inductor with self-resonance at 3 GHz. Ashby et al. [Ashby et al., 1996] used a special process with thick gold metallization to reduce the metallization losses of the spiral realizing a Q of 12 at 3.5 GHz for a 2.8 nH inductor with self-resonance of 10 GHz. Soyuer and Burghartz et al. [Soyuer et al., 1995] proposed using multi-level metallization to realize shunt-connected spirals to effectively thicken the metallization and thus lower the losses. They measured a 2.1 nH inductor with $Q = 9.3$ at 2.4 GHz.

At the IEDM 95 conference, several papers continued this trend, such as Merrill et al. [Merrill et al., 1995] who proposed a series-connected 16.7 nH multi-level inductor with a $Q = 3$. Incidentally, this proposal was also put forward in 1989 at the GaAs symposium by Geen et al. [Geen et al., 1989]. Geen further proposed offsetting or staggering the spirals to reduce the interwinding capacitance. Kim et al. [Kim et al., 1995] demonstrated high-Q inductors fabricated on 10 μm thick polyimide and with 4 μm thick Al metallization achieving a peak Q of 5.5 at 1.2 GHz for a 10 nH inductor. [Burghartz et al., 1995] also presented another multi-level inductor of 1.45 nH achieving a $Q = 24$. In another series of papers [Burghartz et al., 1996b] [Burghartz et al., 1996a], 32 nH and 8.8 nH inductors with $Q = 3$ and $Q = 6.8$ were demonstrated utilizing series and shunt connections of multiple metal layers. In this work, the top metal layer is 2.1 μm of AlCu and resides 10 μm away from a fairly resistive substrate of 12 Ω-cm.

Insulating substrates such as sapphire were demonstrated to yield a Q of 11.9 for a 4 nH inductor [Johnson et al., 1996]. Glass was also used as an insulating substrate by [Dekker et al., 1997] at the IEDM 97 conference yielding inductor values with peak $Q = 40$ at 5 GHz for a 2.6 nH device and $Q = 15$ for a 33.2 nH device at 1.5 GHz. A specially processed lateral high Q coil inductor was demonstrated by [Young et al., 1997] achieving a $Q = 30$ for a 4.8 nH device at 1 GHz.

The work of [Yue and Wong, 1997] [Yue and Wong, 1998] demonstrated electrostatic shielding of inductors from the substrate by utilizing a patterned layer of polysilicon or metal under the device. The patterning is used to minimize eddy currents in the shield, similar to lamination in transformers. The importance of the substrate ground contact placement were illustrated [Burghartz et al., 1997] [Niknejad, 1997]. The use of halo substrate contacts was shown to have a small effect on self-resonant frequency while lowering both the electrical and magnetically induced substrate currents. This is in contrast to the work of [Yue and Wong, 1997] where a patterned ground shield has an adverse effect of self-resonant frequency.

The work of [Park et al., 1997a] [Park et al., 1997b] further illustrated the importance of using a high resistivity substrate (2 $k\Omega$-cm) and thick metal (3.1 μm). This yielded a peak $Q = 17.6$ at 11.25 GHz for a 1.96 nH inductor. The effects of reverse-biasing the substrate through the periphery of the inductor was also investigated in this work. The high substrate resistivity yields depletion depths up to 34 μm for a 5 V reverse bias. [Kim and O, 1997] also illustrated superior self-resonance and Q factor by constructing inductors over reverse-biased wells, dropping the capacitance by a factor of two. Another work utilizing 2 μm thick metal with a high resistivity substrate of 2 $k\Omega$-cm demonstrated a $Q = 12$ at 3 GHz for a 13 nH inductor self-resonating at 7 GHz.

At IEDM 98, a novel buried oxide isolation technique was demonstrated [Erzgraber et al., 1998] to reduce substrate losses. A 2 nH inductor of $Q = 20$ was realized on a 10-20 Ω-cm substrate. Micro-machined solenoid-type coils were demonstrated by [Yoon et al., 1998]. A 2.5 nH inductor with $Q = 19$ at 5.5 GHz occupies 800 μm $\times$ 90 μm, an area comparable to traditional spirals. A 10 nH inductor with $Q = 12.5$ at 2.3 GHz was also illustrated. A shallow junction diffused shield inductor was illustrated by [Yoshitomi et al., 9998]; a $Q = 13$ 2.5 nH inductor was illustrated, an improvement of 80% at 2 GHz. This contrasted with only minor improvements when compared to employing polysilicon shields.

Lee [Lee et al., 1998b] proposed lateral bond wire inductors to realize high-Q and tight tolerance on inductance. This approach differs from using long bond wires [Craninckx and Steyaert, 1995] where manufacturability is a concern. Specifically, a 3.5 nH inductor with $Q = 21$ and self-resonance of 11.3 GHz was illustrated. The tolerance is claimed at better than 5%.

In summary, the experimental evidence suggested clearly that high-Q devices were possible if special processing steps could be added. Furthermore, other works demonstrated that judicious utilization of the metallization layers in a standard CMOS process could yield superior inductors.

3.2 ANALYTICAL RESEARCH

While the above cited works were mainly concerned with achieving high-Q values through processing steps, others have concentrated on studying the inductor loss mechanisms to gain valuable insight in optimizing the geometry and process.

Some early work [Lovelace et al., 1994] employed full *EM* numerical software to analyze the losses in inductors. Others, [Long and Copeland, 1997] [Niknejad and Meyer, 1998] proposed semi-analytical approaches to modeling inductors. The work of [Long and Copeland, 1997] models each spiral segment individually, calculating the inductance using the approach of Greenhouse and Grover, and calculating the capacitance using the 2D approach of [Rabjohn, 1991]. While skin effect was modeled in this approach, proximity effects were not due to a uniform current approximation in the calculation of the partial inductance matrix. Also due to the 2D approach, only rectangular "Manhattan" type geometries can be analyzed, limiting the generality of the approach. Eddy currents in the substrate were neglected and electrically induced currents were calculated indirectly as a free-space Green function was employed. The work of [Niknejad and Meyer, 1998] and this work overcome some of these difficulties by employing a 3D Green function derived over a multi-layer substrate [Niknejad et al., 1998]. Also, current constriction is modeled by sub-dividing the conductor width and thickness using the PEEC formulation. Eddy currents are also treated approximately as we will discuss in Chapter 6.

The current crowding effects of spiral inductors were studied by [Huan-Shang et al., 1997] using the method of moments (MOM) and finite-difference time domain (FDTD) techniques. The work clearly illustrated the importance of modeling the non-uniform current distribution in the spiral inductor, especially in the inner turns where much current constriction occurs due to the high magnetic fields. In fact, [Craninckx and Steyaert, 1997] advocated eliminating the inner turns completely creating a "hollow" spiral to reduce the AC resistance. The work of [Lopez-Villegas et al., 2000] proposes tapering the spiral to minimize the losses of the inner turns.

The work of [Jiang et al., 1997, Hejazi et al., 1998] employs the Green function of circular disks to calculate the distributed capacitance of a circular spiral. The work has limited applicability, though, as a free-space Green function is employed and only applies to circular structures. The work of Rejaei [Rejaei et al., 1998] also derives substrate losses using a circular Green function. This Green function, however, is derived over a conductive substrate and the model is able to predict eddy currents and magnetically induced currents accurately.

The work of Kapur and Long [Kapur and Long, 1997] utilizes an efficient full-wave Green function to calculate the losses over a stratified semi-infinite substrate. The efficiency is gained through efficient numerical techniques. For instance, dense matrices are inverted using the Krylov sub-space approach where matrix-vector produces are computed efficiently using a recursive-SVD algorithm to factor the matrix.

Mohan et al. [Mohan et al., 1999] derive both physical and curve-fit formulas for the inductance value of coils. For the curve-fitting process, extensive *ASITIC* simulations are used (see chapter 7) to derive the coefficients. In another work [Mohan et al., 1998], modeling of lumped transformers is presented.

4. PASSIVE DEVICES ON HIGHLY CONDUCTIVE SI SUBSTRATE

The physically based approaches discussed above completely ignore eddy currents. This was initially not a problem as resistive substrates were employed. However, when the same techniques were applied to a heavily conductive substrate, such as an epi CMOS process, large discrepancies were observed.

This problem was identified by [Craninckx and Steyaert, 1997] through numerical simulation. The simulation was accelerated by assuming 2D rotational symmetry, thus limiting the applicability to circular devices. The work of [Lee et al., 1998a] extended the work presented in [Niknejad and Meyer, 1997] based on eddy current calculations of [Hurley and Duffy, 1995]. This work also suggests that stacked or multi-layer series-connected spirals have superior Q factor when integrated on a highly conductive substrate where eddy current losses dominated. This work also improves upon earlier techniques developed in [Niknejad, 1997] to include such eddy-current induced losses. This will be discussed in Chapter 6.

Even though some of the EM approaches discussed above automatically take eddy currents into account, these techniques are much slower than the techniques presented in this research. As a comparison, simulations with SONNET [Rautio, 1999], a commercially available tool, can run into hours for complicated geometries whereas they take seconds or sub-seconds utilizing the techniques of this book.

Chapter 4

ELECTROMAGNETIC FORMULATION

1. INTRODUCTION

In this chapter we will discuss relevant electromagnetic theorems and assumptions. We begin with Maxwell's equations and derive appropriate scalar and vector potentials. Next we discuss inverting the defining partial differential equations by the Green function technique. We will also discuss various techniques to "hide" the substrate induced losses in Maxwell's equations to simplify the equations as much as possible.

2. MAXWELL'S EQUATIONS
2.1 STATIC SCALAR AND VECTOR POTENTIAL

Many electromagnetic phenomena are closely approximated by Maxwell's partial differential equations over an enormously large range of distances and field strengths [Jackson, 1999]. In the time periodic case, Maxwell's equations become

$$\nabla \cdot \mathbf{B} = 0 \tag{4.1}$$

$$\nabla \cdot \mathbf{D} = \rho \tag{4.2}$$

$$\nabla \times \mathbf{H} = j\omega \mathbf{D} + \mathbf{J} \tag{4.3}$$

$$\nabla \times \mathbf{E} = -j\omega \mathbf{B} \tag{4.4}$$

Assuming that all the materials in question are linear and isotropic, then the number of unknowns in the above equations reduce due to the following constitutive relations

$$\mathbf{D} = \epsilon \mathbf{E} \tag{4.5}$$

$$\mathbf{B} = \mu \mathbf{H} \tag{4.6}$$

$$\mathbf{J} = \sigma \mathbf{E} \tag{4.7}$$

The above relations hold for typical field strengths encountered in microwave engineering. The linearity holds since the internal fields of atoms are orders of magnitude larger than impressed field strengths due to a macroscopic arrangement of conductors.[1]

Note that ϵ, μ, and σ are assumed scalar constants. Assuming that μ is a constant scalar is an excellent approximation for non-magnetic materials. Most magnetic materials, such as paramagnetics and ferromagnetics, though, exhibit both anisotropy, non-linearity, and hysteresis [Kittel, 1996]. A typical IC process, though, only utilizes non-magnetic or weakly diamagnetic materials and thus it is valid to neglect all magnetic properties of the matter in question. On the other hand, Si and most other semiconductors are non-isotropic materials and thus in (4.5) and (4.7) ϵ and σ become tensors. In this work we will neglect this and assume a constant scalar value for ϵ and σ. It should also be noted that both ϵ, μ and σ vary with frequency [Kittel, 1996]. For microwave frequencies of interest, though, we can neglect this variation although its inclusion would not significantly alter the following analysis.

Equations (4.1)–(4.4) can be recast in terms of the electric scalar and magnetic vector potential. From (4.1) it is clear that $\mathbf{B} = \nabla \times \mathbf{A}$ and from (4.4) it follows that [Ramo et al., 1994]

$$\mathbf{E} = -j\omega \mathbf{A} - \nabla \phi \tag{4.8}$$

Using (4.2) and the constitutive relation of (4.5) we have

$$\nabla \cdot \mathbf{E} = -j\omega \nabla \cdot \mathbf{A} - \nabla^2 \phi = \rho/\epsilon \tag{4.9}$$

Since the divergence of $\mathbf{A}$ does not have any physical significance, its value can be chosen arbitrarily. It is common practice to choose its value to simplify the equations as much as possible. For electrostatic problems the Coulomb gauge $\nabla \cdot \mathbf{A} = 0$ is standard but for electromagnetic problems the Lorenz gauge $\nabla \cdot \mathbf{A} = j\omega \mu_0 \epsilon_0 \phi$ is used to simplify Helmholtz's equations [Ramo et al., 1994]. Here, we will invoke the Coulomb gauge whereas in the next section we will examine the full electromagnetic solution. The Coulomb gauge gives us the well-known Poisson's equation

$$\nabla^2 \phi = -\rho/\epsilon \tag{4.10}$$

If we substitute $\mathbf{B} = \nabla \times \mathbf{A}$ into (4.3) we obtain

$$\nabla \times \mathbf{B} = \nabla \times \nabla \times \mathbf{A} = j\omega \mu \epsilon \mathbf{E} + \mu \mathbf{J} \tag{4.11}$$

Using the vector identity

$$\nabla \times \nabla \times \mathbf{A} \equiv \nabla(\nabla \cdot \mathbf{A}) - \nabla^2 \mathbf{A} \tag{4.12}$$

along with the Coulomb gauge and (4.8) results in

$$-\nabla^2 \mathbf{A} = j\omega\mu\epsilon(-j\omega \mathbf{A} - \nabla\phi) + \mu \mathbf{J} \tag{4.13}$$

The first term on the right-hand side of (4.13) results in radiation. For microwave frequencies where the device dimension is much shorter than the resulting wavelength of radiation, this term can be safely neglected. Rearranging we have

$$-(\nabla^2 + \omega^2\mu\epsilon)\mathbf{A} = \mu(-j\omega\epsilon\nabla\phi + \mathbf{J}) \tag{4.14}$$

Consider now the first term of the right hand side of the above equation. Let's denote this term as

$$\mathbf{J}_c = j\omega\epsilon\nabla\phi \tag{4.15}$$

By (4.10) the divergence of the above equation results in

$$\nabla \cdot \mathbf{J}_c = j\omega\epsilon\nabla^2\phi = j\omega\rho \tag{4.16}$$

Now consider the curl of (4.15)

$$\nabla \times \mathbf{J}_c = j\omega\epsilon\nabla \times (\nabla\phi) \equiv 0 \tag{4.17}$$

By the Helmholtz theorem [Collin, 1990], a vector field is uniquely determined up to a scalar constant by its curl and divergence. If we separate the current in (4.14) into two such components we have

$$\mathbf{J} = \mathbf{J}_s + \mathbf{J}_i \tag{4.18}$$

where the subscript s denotes a solenoidal current and the subscript i denotes an irrotational current. Taking the divergence of (4.3) we have

$$\nabla \cdot (\nabla \times \mathbf{B}) \equiv 0 = j\omega\mu\nabla \cdot \mathbf{D} + \mu\nabla \cdot \mathbf{J} \tag{4.19}$$

But since $\nabla \cdot \mathbf{J}_s \equiv 0$ and $\nabla \cdot \mathbf{D} = \rho$ we have

$$\nabla \cdot \mathbf{J}_i = -j\omega\rho \tag{4.20}$$

And by definition $\nabla \times \mathbf{J}_i \equiv 0$. Thus, by Helmholtz's theorem it follows that $\mathbf{J}_i \equiv \mathbf{J}_x$. In other words, (4.14) can be rewritten

$$-\nabla^2 \mathbf{A} = \mu(\mathbf{J} - \mathbf{J}_i) = \mu\mathbf{J}_s \tag{4.21}$$

In summary, Maxwell's equations under electrostatic conditions can be solved by solving the following scalar and vector differential equations

$$\nabla^2\phi = -\rho/\epsilon \tag{4.22}$$

$$\nabla^2\mathbf{A} = -\mu\mathbf{J}_s \tag{4.23}$$

2.2 ELECTROMAGNETIC SCALAR AND VECTOR POTENTIAL

Repeating the above procedure with the Lorenz gauge, $\nabla \cdot \mathbf{A} = -j\omega\mu_0\epsilon_0\phi$, results in the following equations [Ramo et al., 1994]

$$(\nabla^2 + \omega\mu\epsilon)\phi = -\rho/\epsilon \tag{4.24}$$

$$(\nabla^2 + \omega\mu\epsilon)\mathbf{A} = -\mu\mathbf{J} \tag{4.25}$$

For lossy media, the following gauge

$$\nabla \cdot \mathbf{A} = -(j\omega\mu\epsilon + \mu\sigma)\phi \tag{4.26}$$

results in the following equations [Zhou, 1993]

$$(\nabla^2 + \omega\mu\epsilon - j\mu\sigma)\phi = -\rho/\epsilon \tag{4.27}$$

$$(\nabla^2 + \omega\mu\epsilon - j\mu\sigma)\mathbf{A} = -\mu\mathbf{J} \tag{4.28}$$

3. CALCULATING SUBSTRATE INDUCED LOSSES

At microwave frequencies, the electromagnetic fields generated by a passive device penetrate the substrate. For a non-conductive and non-magnetic substrate, such as GaAs, the only significant source of loss inside of the substrate is due to the loss tangent of the material. On the other hand, the conductivity of Si varies considerably from a fairly non-conductive $\rho \sim 10\ k\Omega$-cm for lightly doped Si to fairly conductive at $\rho \sim 10^{-3}\ \Omega$-cm for heavily doped Si. As a result, electromagnetically induced substrate currents flow in the substrate and are a source of loss.

The loss in the Si substrate can be computed from Poynting's theorem. In the time-period case we have [Ramo et al., 1994]

$$\oint_S (\mathbf{E} \times \mathbf{H}^*) \cdot d\mathbf{S} = -\int_V (\mathbf{E} \cdot \mathbf{J}^* + j\omega(\mathbf{H}^* \cdot \mathbf{B} - \mathbf{E} \cdot \mathbf{D}^*))dV \tag{4.29}$$

The surface of the above integration is any surface that completely encloses the Si substrate. The parenthetical term in the above equation represents energy storage and does not lead to loss unless the constitutive relations of (4.5) and (4.6) have imaginary parts. If we ignore the dielectric and magnetic losses of Si in comparison with the conductive losses, only the first term of (4.29) remains. Thus, if we assume that conductive currents dominate so that $\mathbf{J} = \sigma\mathbf{E}$ the integrand simplifies to $\sigma\mathbf{E} \cdot \mathbf{E}^*$.

Now consider making the substitution $\mathbf{J} = \sigma\mathbf{E}$ into (4.3)

$$\nabla \times \mathbf{B} = (j\omega\mu\epsilon + \mu\sigma)\mathbf{E} \tag{4.30}$$

From above, we can define an effective frequency-dependent dielectric constant $\epsilon' = \epsilon + j\sigma/\omega$ and thus remove the current term from Maxwell's equation. Thus, in the calculation of the average loss in (4.29) only the following term remains since $\mathbf{J} \equiv 0$

$$\begin{aligned}-j\omega\mathbf{E}\cdot\mathbf{D}^* &= -j\omega(\epsilon +'\tfrac{j\sigma}{\omega}\mathbf{E}\cdot\mathbf{E}^*) \\ &= -j\omega\epsilon\mathbf{E}\cdot\mathbf{E}^* + \sigma\mathbf{E}\cdot\mathbf{E}^*\end{aligned}$$

Again the first term in the above equation represents the electrical energy stored in the substrate while the second term accounts for the conductive losses.

Thus, we see that to calculate the conductive losses we can simply work with a new system where $\sigma' = 0$ and $\epsilon' = \epsilon + j\sigma/\omega$. So we must now solve

$$\nabla\cdot\mathbf{B} = 0 \tag{4.31}$$

$$\nabla\cdot\mathbf{D} = \rho \tag{4.32}$$

$$\nabla\times\mathbf{B} = j\omega\mu\mathbf{D} \tag{4.33}$$

$$\nabla\times\mathbf{E} = -j\omega\mathbf{B} \tag{4.34}$$

Using operations identical to the previous section we have the modified Poisson's equation

$$\nabla^2\phi = -\rho/\epsilon' \tag{4.35}$$

As we shall see, solving the above equation is equivalent to finding the capacitance matrix of a system of conductors. The introduction of a complex ϵ will make each capacitor lossy by the introduction of a series resistor.

Similarly solving (4.23) in the lossless case leads to the partial inductance matrix. In the case of a conductive substrate, this inductance matrix will become complex where the series loss element includes magnetically induced substrate eddy currents.

Again, performing operations similar to the previous section we obtain

$$\begin{aligned}-\nabla^2\mathbf{A} &= j\omega\mu\epsilon'\mathbf{E} \\ &= j\omega\mu\epsilon'(-j\omega\mathbf{A} - \nabla\phi) \\ &= \omega^2\mu\epsilon'\mathbf{A} - j\omega\mu\epsilon'\nabla\phi \\ &= \omega^2\mu\epsilon\mathbf{A} - j\omega\mu\sigma\mathbf{A} - j\omega\mu\epsilon\nabla\phi - \mu\sigma\nabla\phi\end{aligned} \tag{4.36}$$

Again, neglecting the radiation term we have

$$-\nabla^2\mathbf{A} = \mu(\mathbf{J}_{eddy} - \mathbf{J}_{cond} + \mathbf{J}_{disp}) \tag{4.37}$$

where

$$\mathbf{J}_{eddy} = -j\omega\sigma\mathbf{A} \quad (4.38)$$

$$\mathbf{J}_{disp} = j\omega\epsilon\nabla\phi \quad (4.39)$$

$$\mathbf{J}_{cond} = \sigma\nabla\phi \quad (4.40)$$

Since $\mathbf{J}_{disp}$ represents energy storage and the losses associated with $\mathbf{J}_{cond}$ have been taken into account by solving the modified Poisson's equation (4.22), only the eddy currents contribute loss not accounted for by (4.22). Thus, from a loss perspective we can solve

$$\nabla^2\mathbf{A} = j\omega\mu\sigma\mathbf{A} \quad (4.41)$$

to obtain all the losses associated with the substrate. Note that by the Coulomb gauge the eddy current $\mathbf{J}_{eddy}$ is solenoidal. Furthermore, the displacement current $\mathbf{J}_{disp}$ and conduction currents are irrotational. Thus, the irrotational currents do not give rise to any magnetic effects at low frequencies and neglecting these terms is also justified. In summary, we now solve the following system of equations and accurately account for both the electrically and magnetically induced substrate losses

$$\nabla^2\mathbf{A} = j\omega\mu\sigma\mathbf{A} + \mu\mathbf{J}_{src} \quad (4.42)$$

$$\nabla^2\phi = -\frac{\rho}{\epsilon + j\sigma/\omega} \quad (4.43)$$

The only loss mechanism that we have neglected is radiation.

4. INVERSION OF MAXWELL'S DIFFERENTIAL EQUATIONS

Given a fixed current and charge "source" density distribution, one can solve Maxwell's equation in closed form. This is achieved by inverting the partial differential equations of (4.22) and (4.23). In free-space it is well known that such a solution is given by

$$\mathbf{A}(\mathbf{r}) = \mu \int_V \frac{\mathbf{J}(\mathbf{r}')e^{-jkR}dV'}{4\pi R} \quad (4.44)$$

$$\phi(\mathbf{r}) = \int_V \frac{\rho(\mathbf{r}')e^{-jkR}dV'}{4\epsilon\pi R} \quad (4.45)$$

where $R = |\mathbf{r} - \mathbf{r}'|$. This can be generalized by the introduction of the Green function. Any non-singular linear differential operator $\mathcal{L}$ subject to boundary conditions can be inverted by application of the appropriate Green function [Roach, 1982].

$$x = \mathcal{L}^{-1}y = \int G(x, x')y(x')dx' \quad (4.46)$$

where the function G is the solution to the following problem

$$\mathcal{L}G(x, x') = \delta(x - x') \qquad (4.47)$$

and G also satisfies the boundary conditions. Formally, to see that this is indeed the solution simply observe that it does satisfy the original equation

$$\mathcal{L}\int G(x, x')y(x')dx' = \int \mathcal{L}G(x, x')y(x')dx' = \int \delta(x-x')y(x')dx' = y(x) \qquad (4.48)$$

where we have interchanged the order of the operator and the integration by invoking linearity. We have also used the "sifting" property of the Dirac delta function. To prove the above result in rigorously is beyond the scope of this book and such a proof can be found elsewhere. This result is of course intuitive when (4.46) is interpreted as linear superposition integral. In fact, the electrical engineer is already familiar with the Green function solution to time-domain ordinary differential equations [2].

For the case of scalar and vector partial differential equations, we can obtain the result by invoking Green's first and second theorems

$$\int_V (f_1 \nabla^2 f_2 + \nabla \mathbf{f_1} \cdot \nabla \mathbf{f_2})dV = \oint_S f_1 \nabla f_2 \cdot d\mathbf{S} \qquad (4.49)$$

$$\int_V (f_1 \nabla^2 f_2 - f_2 \nabla^2 f_1)dV = \int_S (f \nabla \mathbf{f_2} - f_2 \nabla \mathbf{f_1}) \cdot d\mathbf{S} \qquad (4.50)$$

Applying the second identity to Poisson's equation with $f_1 = \phi$ and $f_2 = G$ we obtain

$$\int_V (G\nabla^2 \phi - \phi \nabla^2 G)dV = \int_S (G\nabla \phi - \phi \nabla \mathbf{G}) \cdot d\mathbf{S} \qquad (4.51)$$

And noting that $\nabla^2 \phi = -\rho/\epsilon$ and $\nabla^2 G = \delta(\mathbf{r} - \mathbf{r}')$ we obtain

$$\phi = \frac{-1}{\epsilon}\int_V G\rho dV + \int_S \int_S (\phi \nabla \mathbf{G} - G\nabla \phi) \cdot d\mathbf{S} \qquad (4.52)$$

We can simplify the above equation if we choose zero potential on the surface S and the second term disappears. Working with the vector form of Green's second identity

$$\int_V (\mathbf{F_1} \cdot \nabla \times \nabla \times \mathbf{F_2} - \mathbf{F_2} \cdot \nabla \times \nabla \times \mathbf{F_1})dV = \int_S (\mathbf{F_2} \times \nabla \times \mathbf{F_1} - \mathbf{F_1} \times \nabla \times \mathbf{F_2}) \cdot d\mathbf{S} \qquad (4.53)$$

and let $\mathbf{G}$ be the solution to

$$\nabla \times \nabla \times \mathbf{G} = -\mu \delta(\mathbf{r} - \mathbf{r}')\hat{m} \qquad (4.54)$$

Solving the above equation in turn for $\hat{m} = \hat{x}$, $\hat{m} = \hat{y}$, and $\hat{m} = \hat{z}$, results in three vector functions $\mathbf{G_{x,y,z}}$. For an isotropic medium such as free-space it follows that $\mathbf{G_x} = \mathbf{G_y} = \mathbf{G_z} = \mathbf{G}$. Now if we define a dyadic quantity $\bar{\mathbf{G}} = \mathbf{G_x}\hat{x} + \mathbf{G_y}\hat{y} + \mathbf{G_z}\hat{z}$ it can then be shown that for an arbitrary current distribution the solution to the vector potential is given by [Collin, 1990]

$$\mathbf{A}(\mathbf{r}) = \int_V \bar{\mathbf{G}}(\mathbf{r},\mathbf{r}') \cdot \mathbf{J}(\mathbf{r}')dV \tag{4.55}$$

where the "scalar" product $\bar{\mathbf{G}}(\mathbf{r},\mathbf{r}') \cdot \mathbf{J}(\mathbf{r}')$ results in

$$\mathbf{G_x} \cdot \mathbf{J}_x + \mathbf{G_y} \cdot \mathbf{J}_y + \mathbf{G_z} \cdot \mathbf{J}_z \tag{4.56}$$

5. NUMERICAL SOLUTIONS OF ELECTROMAGNETIC FIELDS

Many diverse techniques exist for obtaining numerical solutions of (4.1)–(4.4). Most techniques can be categorized into those that discretize the fields E and B, the so-called domain techniques, versus techniques that discretize the sources, the so-called boundary methods, versus techniques that discretize the continuous differential or integral operators. Techniques which discretize the fields are appropriate when field solutions are desired in a complicated non-uniform volume. This is the situation, for example, for device simulators where a complicated doping profile due to diffusion creates regions of varying conductivity both laterally and vertically. One such approach is the Finite Difference Equation (FDE) techniques which discretize the partial differential equations of Helmholtz which are obtained by well-known transformation of (4.1)–(4.4). Note that this technique approximates a continuous operator by a discrete one. Another popular technique, the Finite Element Method (FEM), discretizes the equivalent functional relations derived by the techniques of calculus of variations. Here one approximates the field quantity, not the operator. The FEM technique, like all variational techniques, is related to Green's first identity

$$(\mathcal{L}u, u) = (\mathcal{L}_{-1}u, \mathcal{L}_{-1}u) + B(u, u) \tag{4.57}$$

If the operator $\mathcal{L}$ is self-adjoint and positive definite, then the solution of $\mathcal{L}u = f$ minimizes the following functional

$$J(u) = (\mathcal{L}u, u) - 2(f, u) \tag{4.58}$$

Other techniques, the ones pursued in this work, result from Green's second identity

$$(\mathcal{L}u, v) - (\mathcal{L}^*v, u) = B(u, v) \tag{4.59}$$

We discretize the sources of the fields by the method of moments (MoM). This results in considerable savings in the number of unknowns for typical microwave frequencies. This happens because charge accumulates at the surface of conductors and thus only a shallow thickness of conductors need to be discretized. In other words, only the charge and current along the surface of conductors needs discretization as opposed to the entire volume of the problem at hand. Of course, FEM and FDE techniques can employ non-uniform discretization to capture rapid variations of the fields at the surfaces of conductors to reduce the number of unknowns considerably. The conducting substrate presents a problem for such techniques, considerably increasing the number of "unknowns" in the equations, but application of an appropriate Green function eliminates the need to discretize the substrate.

6. DISCRETIZATION OF MAXWELL'S EQUATIONS

We begin by neglecting displacement current in the volume of good conductors. The electric field is thus related to current by

$$\mathbf{J} = \sigma \mathbf{E} = \sigma(-\nabla\phi - j\omega A) \tag{4.60}$$

Note that we cannot ignore displacement currents in the substrate so we will instead only speak of conductive currents in conductors. The substrate currents appear only implicitly in the equations below through the action of the Green function

$$\mathbf{A} = \mu \int \bar{\mathbf{G}} \cdot \mathbf{J} dV \tag{4.61}$$

Substituting the above equation in the previous we have

$$\frac{\mathbf{J}}{\sigma} + j\omega\mu \int \bar{\mathbf{G}} \cdot \mathbf{J} dV = -\nabla\phi \tag{4.62}$$

The above equation is coupled to Poisson's equation by

$$\nabla^2 \phi = -\frac{\rho}{\epsilon} \tag{4.63}$$

To solve this equation numerically, we make the following assumptions:

- Displacement current is neglected inside volume element of conductors.
- The currents and charges in volume elements can be considered uniform if the volume elements are made sufficiently small.
- The potential does not vary appreciably in a volume element.
- The effects of substrate charges are captured by the electric Green functions averaged over the volume of conductors.

- The effects of substrate currents are captured by a scalar Green function averaged over the volume of conductors. The scalar nature is obtained by neglecting z-directed current coupling and by assuming radial symmetry laterally.

- Proper boundary conditions can be imposed such that the surface terms in the scalar and vector potentials vanish.

Thus, we discretize the source current and charge

$$\mathbf{J}(\mathbf{r}) = \sum \frac{I_k}{V_k} \psi_k(\mathbf{r}) \hat{m}_k \tag{4.64}$$

Where we assume that the set $\{\psi_k\}$ spans the space. Since the underlying equations are self-adjoint, we can always choose an orthonormal set [Roach, 1982]

$$<\psi_i, \psi_j> = \delta_{ij} \tag{4.65}$$

where each function ψ_i is an eigenfunction of the underlying operator. In such a case, great numerical accuracy can be obtained by only retaining a few terms in (4.64). We can also show that this case leads to uniform convergence of (4.64). However, this choice will lead to difficult numerical integrations. Another choice is spatially localized functions that automatically satisfy the orthogonality conditions of (4.65). The simplest of such functions are piecewise continuous

$$\psi_i(\mathbf{r}) = \begin{cases} 1 & \text{if } \mathbf{r} \in V_i \\ 0 & \text{otherwise} \end{cases} \tag{4.66}$$

With such a choice we have the following equivalent matrix equation [Ruehli, 1974] [Kamon et al., 1994a]

$$R_k I_k + j\omega \sum_j M_{jk} I_j = \Delta \phi_k \tag{4.67}$$

where we compute M by

$$M_{jk} = \frac{\mu}{A_j A_k} \int_{V_k} \int_{V_j} G_M(\mathbf{r}, \mathbf{r}') dV' dV \tag{4.68}$$

Let $\check{Z} = R + M$ be the impedance matrix such that we can write

$$\check{\mathbf{v}} = \Delta \phi = \check{Z} \check{\mathbf{i}} \tag{4.69}$$

We can immediately interpret the above results physically by noting that the moments in (4.64) are physical currents, M is related to the partial inductance matrix and the diagonal matrix R is simply the DC resistance of each conductor [Ruehli, 1974]. The magnetic Green function G_M is no longer a scalar function

Electromagnetic Formulation 55

due to the presence of the substrate. If we assume a substrate is infinite in extent and uniform laterally, and stratified vertically, as shown in Fig. 6.1, then for lateral currents G_M is effectively a scalar. In the following analysis we will make this assumption since IC processes employ metal layers parallel to the surface of the substrate. Vertical currents flow through vias but we will ignore the substrate coupling between such via currents and the substrate.

Now if we choose the volume elements V_i sufficiently small that the current within this volume is nearly uniform, this approximation will be valid. Unfortunately, the higher the frequency the smaller we will need to make the volume, leading to more terms in the summation of (4.64). We can ameliorate this somewhat by choosing non-uniform volume elements V_i since currents will tend to concentrate at the "skin" of conductors whereas currents deep inside conductors will approach a more uniform and smaller value.

Treating the charges in the system in a similar manner, we expand the charge density ρ into a set of localized functions

$$\rho(\mathbf{r}) = \sum \frac{q_k}{U_k} \zeta_k(\mathbf{r}) \tag{4.70}$$

At this point we choose not to discretize charge and current necessarily in the same manner

$$\zeta_i(\mathbf{r}) = \begin{cases} 1 & \text{if } \mathbf{r} \in U_i \\ 0 & \text{otherwise} \end{cases} \tag{4.71}$$

However, at some level we must impose charge conservation

$$\nabla \cdot \mathbf{J} = -\frac{d\rho}{dt} \tag{4.72}$$

And this requires some consistency in choosing the boundaries of each volume element U and V. Note that the charge in the substrate, similar to the substrate currents, are treated implicitly by the Green function. Inverting Poisson's equation we have

$$\phi(\mathbf{r}) = \int_V G_E(\mathbf{r}, \mathbf{r}') \rho(\mathbf{r}') dV' \tag{4.73}$$

The average potential in some volume U_k can be obtained by

$$\tilde{v}_k = \frac{1}{U_k} <\zeta_k, \phi(\mathbf{r})> = \frac{1}{U_k} \int_{U_k} \phi(\mathbf{r}) dV \tag{4.74}$$

and expanding the expression for ϕ

$$\tilde{v}_k = \frac{1}{U_k} \int_{U_k} \int_V G(\mathbf{r}, \mathbf{r}') \rho(\mathbf{r}) dV' dV \tag{4.75}$$

and further expanding the expression for ρ and interchanging the order of the summation and integration

$$\tilde{v}_k = \frac{1}{U_k} \int_{U_k} \sum_j \frac{q_j}{U_j} \int_V G(\mathbf{r},\mathbf{r}')\zeta_j dV' dV \qquad (4.76)$$

The innermost volume integral is localized to U_j due to the locality of ζ. Changing the order of the summation and integration once more

$$\tilde{v}_k = \frac{1}{U_k} \sum_j \frac{q_j}{U_j} \int_{U_k} \int_{U_j} G(\mathbf{r},\mathbf{r}') dV' dV \qquad (4.77)$$

We thus have the following matrix equation

$$\tilde{\mathbf{v}} = P\tilde{\mathbf{q}} \qquad (4.78)$$

where each matrix element of P is given by

$$P_{jk} = \frac{1}{U_k U_j} \int_{U_k} \int_{U_j} G(\mathbf{r},\mathbf{r}') dV dV' \qquad (4.79)$$

In summary, we have converted the difficult coupled integro-differential equations (4.62,4.63) to the following matrix equations

$$\check{\mathbf{v}} = \check{Z}\check{\mathbf{i}} \qquad (4.80)$$
$$\tilde{\mathbf{v}} = P\tilde{\mathbf{q}} \qquad (4.81)$$

Note that elements of $\check{\mathbf{v}}$ and $\check{\mathbf{i}}$ represent the current and voltage in the volume space of the set V and the elements of $\tilde{\mathbf{v}}$ and $\tilde{\mathbf{q}}$ are the voltages and charges defined in the volume space of the set U. Let $\mathbf{v}$ and $\mathbf{i}$ be vectors defined in the set W, a more coarse volume space than V and U. Thus for any element U_k in the set U, we have $U_k \subset W_j$ for some j. Also, we require that $W_j = \sum_{k \in K} U_k$ for some finite set K. Similar restrictions apply in the relation of W to V. Due to the conservation of charge (KCL), the elements of i are related to the elements of $\tilde{i}$ as follows

$$i_j = \sum_{\forall k, V_k \in W_j} \tilde{i}_k \qquad (4.82)$$

In matrix notation we thus have

$$\mathbf{i} = S\tilde{\mathbf{i}} \qquad (4.83)$$

where the ith row of matrix S has unity terms corresponding to sub-elements and zeros elsewhere. For simplicity, we also assume that all the voltages in the sub-space of V are equal. In matrix notation this becomes

$$\check{\mathbf{v}} = S^T \mathbf{v} \qquad (4.84)$$

Combining the above results we obtain

$$\mathbf{v} = Z^M \mathbf{i} \qquad (4.85)$$

where

$$Z^M = (S\tilde{Z}^{-1}S^T)^{-1} \qquad (4.86)$$

A similar argument applied to the charges gives us

$$\mathbf{q} = \Sigma \tilde{\mathbf{q}} \qquad (4.87)$$

and

$$\tilde{\mathbf{v}} = \Sigma^T \mathbf{v} \qquad (4.88)$$

where Σ plays the same role as S. Thus we have

$$q = (\Sigma P^{-1} \Sigma^T) v = Z^C v \qquad (4.89)$$

In summary, we have reduced Maxwell's equations to discrete currents i, voltages v, and charges q which are related by two matrices, Z^C and Z^M. These matrices are in fact the partial inductance and capacitance matrix. Thus, in place of Maxwell's equations we may work with equivalent lumped circuits. Ruehli [Ruehli, 1974] has shown that this procedure is equivalent to solving Maxwell's equation.

Notes

1. Modern IC processes, though, are capable of producing very thin and small structures leading to very large electric fields. The gate oxide of a modern CMOS process, for instance, can be tens of angstroms thick, involving perhaps tens or hundreds of atoms.
2. The Impulse Response Function $H(t, \tau)$.

Chapter 5

INDUCTANCE CALCULATIONS

1. INTRODUCTION

In this chapter we will define inductance, calculate the low frequency inductance of several common configurations of conductors, and extend the results of our calculations to high frequency.

Up until recently, the calculation of inductance has received relatively little attention from the integrated circuit community. Only very specialized fields, such as power engineering, have dealt directly with inductors. The reason for this is simply related to the relatively minor role that inductors have played at lower frequencies, especially in the IC environment. In such cases, parasitic capacitance plays a dominant role in determining circuit behavior whereas the effects of parasitic inductance are minor. The physical origin of this stems from the relatively small size of integrated circuits. The typical inductance associated with long integrated circuit metal traces is on the order of 1 nH, a small enough value of inductance as to produce negligible reactance at typical IC frequencies less than 100 MHz. Today, RF, microwave, and digital ICs are operating at 1 - 10 GHz and these inductive effects are no longer negligible.

Comparatively, magnetic forces tend to have longer range effects than electric forces. This can be attributed to the absence of magnetic monopoles. Whereas electric field lines terminate on charges in nearby conductors and the substrate, magnetic field lines are divergenceless and only wane in the presence of induced eddy currents.

At very high frequencies, frequencies often encountered by the microwave community, the skin depth δ is small and magnetic fields are confined to the external volume of conductors. In such a case, the inductances of a system of coupled multi-conductor transmission lines can be derived from the capacitance matrix. Therefore, more emphasis has been placed on deriving the capacitance matrix as the source of capacitance is charge derived from a scalar Poisson's equation whereas the source of inductance is from currents satisfying the vector Poisson's equation.

The situation has changed as on-chip frequencies have increased to the point where magnetic effects are appreciable. In his pioneering work, Ruehli [Ruehli, 1972] focused attention on efficient techniques for predicting inductive effects in an integrated circuit environment at moderate frequencies where internal inductance cannot be ignored. These techniques are widely applicable, even to board and package environments.

2. DEFINITION OF INDUCTANCE

Inductance is the electric dual of capacitance. While capacitors store electric energy, inductors store magnetic energy. The self-inductance of a circuit can be thought of as the "mass", or resistance to change in "motion", of the circuit. The origin of this "mass" comes from the magnetic field of a circuit, which originates from the currents in the circuit. Any change in the current of a circuit induces a change in the magnetic field. From Faraday's law we know that a changing magnetic field induces an electric field. From Lenz's law we can deduce further that this induced electric field always opposes further change in the current.

2.1 ENERGY DEFINITION

From circuit theory the total magnetic energy stored by an inductor is given by

$$W_m = \frac{1}{2}LI^2 \tag{5.1}$$

If we thus calculate the total magnetic energy in a volume V by a physical inductor

$$W_m = \frac{1}{2}\int_V \mathbf{B} \cdot \mathbf{H} dV \tag{5.2}$$

we can equate the above quantities to obtain the inductance.

We have already alluded to the energy definition of an inductor in Chapter 2 when we examined the complex power flowing into a black box

$$P = \frac{1}{2}\oint_S \mathbf{E} \times \mathbf{H}^* \cdot d\mathbf{s} = P_l + 2j\omega(W_m - W_e) \tag{5.3}$$

implying that the device acts inductively if $W_m > W_e$. This also has some interesting implications for the impedance of a lossless passive device as a function of frequency. It can be shown that [Ramo et al., 1994]

$$\frac{dX}{d\omega} = 4\frac{W_m + W_e}{II^*} \tag{5.4}$$

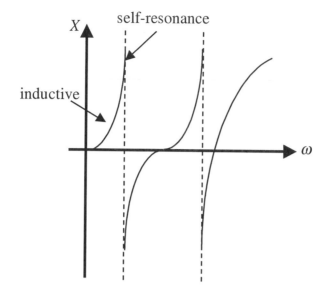

Figure 5.1. The reactance of a lossless passive device as a function of frequency.

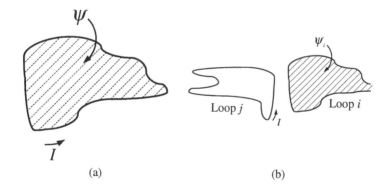

Figure 5.2. (a) An isolated current loop. (b) A magnetically coupled pair of loops. Current only flows in loop j.

which implies that the slope of the impedance versus frequency is never negative. Such a typical profile as a function of frequency is shown in Fig. 5.1.

2.2 MAGNETIC FLUX OF A CIRCUIT

Consider an arbitrary closed-circuit formed by conductors, as shown in Fig. 5.2a. The magnetic flux of this circuit is defined as the magnetic field crossing the cross-sectional area of the circuit

$$\psi = \oint_S \mathbf{B} \cdot d\mathbf{S} \tag{5.5}$$

The origin of the magnetic field $\mathbf{B}$ is from the circuit itself since there are no other currents in the system. Now, the self-inductance of the circuit is simply defined as

$$L = \frac{\psi}{I} \tag{5.6}$$

where I is the total current flowing in the circuit. If we now consider an arrangement of loops as shown in Fig. 5.2b, then if we let current flow in loop j and measure the impinging flux on loop i, we have the following definition of mutual inductance

$$M_{ij} = \frac{\psi_i}{I_j} \tag{5.7}$$

where

$$\psi_i = \oint_{S_i} \mathbf{B} \cdot d\mathbf{S} \bigg|_{I_k = 0 \ \forall \ k \neq j} \tag{5.8}$$

By Faraday's well-known law, the voltage induced on a loop is related to the flux as follows

$$V = \frac{d\psi}{dt} \tag{5.9}$$

and by the definition of inductance this is simply

$$V = L\frac{dI}{dt} \tag{5.10}$$

2.3 MAGNETIC VECTOR POTENTIAL

From Maxwell's equation, we know that all physically observable magnetic fields are solenoidal, i.e. $\nabla \cdot \mathbf{B} = 0$ and thus we can write $\mathbf{B} = \nabla \times \mathbf{A}$ for some vector potential function $\mathbf{A}$. Note that we are free to choose the divergence of $\mathbf{A}$ as we wish. From the definition of flux, we can invoke Stoke's theorem to obtain

$$\psi = \oint_S (\nabla \times \mathbf{A}) \cdot d\mathbf{S} = \oint_C \mathbf{A} \cdot dl \tag{5.11}$$

In many practical cases, it is more convenient to work with the magnetic vector potential $\mathbf{A}$ than the magnetic field $\mathbf{B}$. Although the magnetic field has a more physical interpretation, the magnetic vector potential is often simpler to derive[1]. The governing equation for the vector potential is the vector Poisson's equation derived in Chapter 4

$$\nabla^2 \mathbf{A} = \mu \mathbf{J} \tag{5.12}$$

The solution of this equation can be written with the aid of a dyadic Green function $\bar{\mathbf{G}}$

$$\mathbf{A}(\mathbf{r}) = \mu \int_{V'} \bar{\mathbf{G}}(\mathbf{r},\mathbf{r}') \cdot \mathbf{J}(\mathbf{r}')dV' \tag{5.13}$$

where for instance the component of $\bar{\mathbf{G}}$ for a test source in the direction of $\hat{x}_i$ is a solution of

$$\nabla^2 \mathbf{G_i} = \mu\delta(\mathbf{r}-\mathbf{r}')\hat{x}_i \tag{5.14}$$

In an isotropic medium, such as free-space or a spherically symmetric arrangement of conductors, $G_x = G_y = G_z$ and $G_{km} = 0 \; \forall \; k \neq m$ and thus the equation can be simplified

$$\mathbf{A}(\mathbf{r}) = \mu \int_{V'} G(\mathbf{r},\mathbf{r}')\mathbf{J}(\mathbf{r}')dV' \tag{5.15}$$

By definition, then, the mutual inductance between two loops can be derived from the voltage induced in one loop (unprimed coordinates) for a changing current flowing in some "source" loop (primed coordinates). This voltage is given by

$$V_{\text{fld}} = -\oint_C \mathbf{E} \cdot d\mathbf{l} \tag{5.16}$$

From Maxwell's equation, the electric field is given by

$$\mathbf{E} = -\nabla\phi - \frac{d\mathbf{A}}{dt} \tag{5.17}$$

Since the first term is "conservative" in nature, its line integral along a closed path yields zero. Thus, we have

$$V_{\text{fld}} = \frac{d}{dt}\oint_C \mathbf{A} \cdot d\mathbf{l} = M\frac{dI}{dt} \tag{5.18}$$

Hence

$$M = \frac{1}{I}\oint_C (\mu \int_{V'} G(\mathbf{r},\mathbf{r}')\mathbf{J}(\mathbf{r}')dV') \cdot d\mathbf{l} \tag{5.19}$$

At low frequencies, the current distribution $\mathbf{J}$ is uniform and the second integration therefore only involves the function G. We can view this volume integral as the line integral of the cross-sectional area of the conductor along the path of the conductor. Therefore

$$\int_{V'} G(\mathbf{r},\mathbf{r}')\mathbf{J}(\mathbf{r}')dV' = I\oint_{C'}(\int_{A'} G(\mathbf{r},\mathbf{r}')dA')d\mathbf{l}' = I\oint_{C'}\tilde{G}(\mathbf{r},\mathbf{r}')d\mathbf{l}' \tag{5.20}$$

where $\tilde{G}$ is the value of G averaged over the cross-sectional area of the "source" circuit at any location along the loop. Substituting the above in the equation for inductance we have

$$M = \mu \oint_C \oint_{C'} \tilde{G}(\mathbf{r},\mathbf{r}')d\mathbf{l} \cdot d\mathbf{l}' \tag{5.21}$$

The current factors out in the above equation and thus we see that the inductance at low frequency is simply related to the geometry of conductors in question. For the case of free-space

$$G(\mathbf{r}, \mathbf{r}') = \frac{1}{4\pi |\mathbf{r} - \mathbf{r}'|} \qquad (5.22)$$

If the source loop is filamental, then we have Neumann's famous result [Ramo et al., 1994]

$$M = \frac{\mu}{4\pi} \oint \oint \frac{d\mathbf{l} \cdot d\mathbf{l}'}{R} \qquad (5.23)$$

which incidentally proves the reciprocity relation $M_{ij} = M_{ji}$.

Evidently, the above result is dependent on the path of integration in the "source" loop. For instance, if we take the inner most turn in a loop of finite width, the value of inductance is different from the outer most loop, due to path length difference and the non-uniformity of the magnetic field. This path dependence is in fact a paramount property of a solenoidal field. Thus, to obtain a path independent mutual inductance, we should average M over all paths. Consider the set of all paths parameterized by $s \in [0, 1]$ tangent to the current flow $K = \{\mathbf{r}_i(s) : \mathbf{n}(s) \cdot \mathbf{J} = 0\}$. The vector $\mathbf{r}_i(s)$ traces one particular path and the vector $\mathbf{n}(s)$ is always normal to the path. Let the average path be described by the following convergent infinite sum

$$\mathbf{r}(s) = \sum_{j \in K} \mathbf{r}_j(s) \qquad (5.24)$$

and let this path be described by C_1. Now let us integrate the field quantity $\mathbf{E}'$ along this path to obtain a scalar value with units of voltage

$$V' = -\oint_{C_1} \mathbf{E}' \cdot d\mathbf{l}_1 \qquad (5.25)$$

where the field quantity $\mathbf{E}'$ is the average value of the electric field along the cross-section of the path C_1

$$\mathbf{E}'(r) = \int_{A_1} \mathbf{E}(\mathbf{r}) dA_1 \qquad (5.26)$$

We can thus interpret (5.25) as the average voltage induced in our "source" loop. Thus, the average mutual inductance becomes

$$M = \mu \oint_{C_1} \oint_{C'} H(\mathbf{r}, \mathbf{r}') d\mathbf{l} \cdot d\mathbf{l}' \qquad (5.27)$$

where H is given by

$$H(\mathbf{r}, \mathbf{r}') = \int_{A_{\text{src}}} \int_{A_{\text{fld}}} G(\mathbf{r}, \mathbf{r}') dA_{\text{src}} dA_{\text{fld}} \qquad (5.28)$$

We can interpret the function $H(\mathbf{r},\mathbf{r}')$ as the average value of G evaluated for all points in the cross-sectional areas of the source and field loops at a given position $\mathbf{r}$ in the field and $\mathbf{r}'$ in the source loops.

3. PARALLEL AND SERIES INDUCTORS

When N inductors are connected in series, the effective mutual inductance can be computed easily since the voltage across the series connection gives

$$V = \sum_{i=1}^{N} V_i = \sum_{i=1}^{N}\sum_{j=1}^{N} sM_{ij}I_j = Is\sum_{i=1}^{N}\sum_{j=1}^{N} M_{ij} \qquad (5.29)$$

where we have assumed that the branch current is equal to I for all branches since we are neglecting displacement current at low frequencies. Clearly, then

$$L_{se} = \sum_{i=1}^{N} L_i + 2\sum_{i=1}^{N}\sum_{j\neq i} M_{ij} \qquad (5.30)$$

The case of parallel connected inductors is more complicated. The result has already been stated in Chapter 2

$$L_{sh} = \frac{1}{\sum_{i=1,j=1}^{N} K_{ij}} \qquad (5.31)$$

where the matrix K is the inverse of the partial inductance matrix M. To see that this is true in general, consider N-shunt connected coupled inductors described by the partial inductance matrix M,

$$\mathbf{v} = j\omega M \mathbf{i} \qquad (5.32)$$

where element v_j of vector $\mathbf{v}$ is the voltage across jth inductor and the element i_j of vector $\mathbf{i}$ is the current through the jth inductor. Since the voltages are connected in shunt, all voltages are equal

$$v_x = v_1 = v_2 = \cdots = v_N \qquad (5.33)$$

and thus

$$\mathbf{v} = v_x \mathbf{s} \qquad (5.34)$$

where $\mathbf{s} = \begin{pmatrix} 1 \\ \vdots \\ 1 \end{pmatrix}$ and the total current is given by

$$i_x = i_1 + i_2 + \cdots + i_N = \mathbf{s}^T \mathbf{i} \qquad (5.35)$$

Since the effective input inductance is given by

$$L_{sh} = \frac{v_x}{j\omega i_x} \quad (5.36)$$

we have from (5.32)

$$(j\omega M)^{-1}\mathbf{v} = (j\omega M)^{-1}\mathbf{s}v_x \quad (5.37)$$

and further

$$\mathbf{s}^T(j\omega M)^{-1}\mathbf{s}v_x = \mathbf{s}^T\mathbf{i} = i_x \quad (5.38)$$

and

$$L_{sh} = \frac{1}{\mathbf{s}^T M^{-1}\mathbf{s}} \quad (5.39)$$

which establishes (5.31). In [Ruehli, 1972] the above result is stated in the following form

$$L_{sh} = \left[2\sum_{i=1}^{N-1}\sum_{j=i+1}^{N} \operatorname{cof} M_{ij} + \sum_{i=1}^{N} \operatorname{cof} M_{ii} \right]^{-1} \det M \quad (5.40)$$

4. FILAMENTAL INDUCTANCE FORMULAE FOR COMMON CONFIGURATIONS

Consider a $\hat{\mathbf{z}}$ directed filament of finite length l carrying a current I in free-space located at the origin of a rectangular coordinate system. The magnetic vector potential at any point in space is given by (5.15)

$$\mathbf{A}(x,y,z) = \hat{\mathbf{z}}\frac{\mu_0 I}{4\pi}\int_{-l/2}^{l/2} \frac{1}{\sqrt{x^2+y^2+(z-z')^2}}dz' \quad (5.41)$$

Performing the above integration we have

$$A_z(x,y,z) = \hat{\mathbf{z}}\frac{\mu_0 I}{4\pi}\log\left[\frac{(z+l/2)+\sqrt{x^2+y^2+(z+l/2)^2}}{(z-l/2)+\sqrt{x^2+y^2+(z-l/2)^2}}\right] \quad (5.42)$$

Now consider the mutual magnetic inductance of two such parallel filaments separated by a distance d. If we integrate the expression for A_z along the path of the second filament we obtain

$$M(l,d) = \int_{-l/2}^{l/2} A_z(d,0,z')dz' \quad (5.43)$$

The above integral can be simplified into the following form

$$M(l,d) = \frac{\mu_0}{4\pi}l\left[\log\left(\sqrt{1+\left(\frac{l}{d}\right)^2}+\frac{l}{d}\right) - \sqrt{1+\left(\frac{d}{l}\right)^2}+\frac{d}{l}\right] \quad (5.44)$$

The above expression is the starting point for calculating the mutual inductance between two arbitrary parallel filaments. The technique is described in Grover [Grover, 1946] and Ruehli [Ruehli, 1972].

5. CALCULATION OF SELF AND MUTUAL INDUCTANCE FOR CONDUCTORS

Following the averaging procedure described in (5.21), we can find the mutual inductance between two parallel conductors of arbitrary cross-section by integrating the filamental formula (5.44) over the area of the cross-section of the conductors

$$M_{12} = \frac{1}{A_1 A_2} \int_{A_1} \int_{A_2} M(L,r) dA_1 dA_2 \qquad (5.45)$$

where $r = \sqrt{(x_1 - x_2)^2 + (y_1 - y_2)^2 + (z_1 - z_2)^2}$. For the case of self-inductance, we can set $A_1 = A_2$ and perform the integration. This will result in an integrable singularity in the calculation.

5.1 THE GEOMETRIC MEAN DISTANCE (GMD) APPROXIMATION

If the filaments under consideration are long such that $L \gg d$, then the results for $M(L,d)$ can be simplified if we neglect the term $\left(\frac{d}{L}\right)^2$

$$M(L,d) \approx \frac{\mu_0}{4\pi} L \left(\frac{d}{L} - \log d + \log 2L - 1\right) \qquad (5.46)$$

Integration of the above expression across the cross-section yields a simpler calculation. The only difficult term to integrate is the $\log(d)$. This term, the *Geometric Mean Distance*, or GMD, is the average value of the logarithm of the distance between all points between the conductors. This integral can be calculated for several simple configurations, especially for arbitrary rectangular cross-sections. Some of these results can be found in [Niknejad, 1997].

A simple interpretation of the above result can be obtained if we make the $L \gg d$ approximation from the outset. If we let $L \to \infty$ we have the following 2D magnetic vector potential

$$\mathbf{A}(x,y) = \hat{z} \frac{\mu_0 I}{4\pi} \log \sqrt{x^2 + y^2} \qquad (5.47)$$

And thus the mutual inductance per unit length between two infinite parallel conductors is given by

$$M_{12} = \frac{\mu_0}{4\pi} \int_{A_1} \int_{A_2} \log(r) dA_1 dA_2 \qquad (5.48)$$

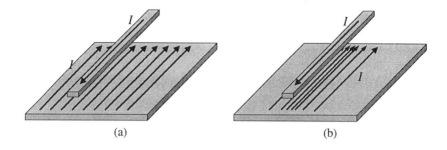

Figure 5.3. (a) Ground current at low frequency. (b) Ground current at high frequency.

where r takes on the value of the distance between every pair of points in the cross-section of each conductor. With the exception of the Arithmetic Mean Distance (AMD) correction factor, this is exactly the same expression found with the GMD approximation.

6. HIGH FREQUENCY INDUCTANCE CALCULATION
6.1 BACKGROUND

All the equations derived thus far for inductance have made the assumption of uniform current distribution within the volume of conductors. While this applies to low frequencies, we know that at high frequencies currents redistribute to minimize the energy of the system. Consider, for instance, a conductor carrying a low frequency current above a ground plane as shown Fig. 5.3a. At low frequencies (strictly DC) the return current flowing in the ground plane flows uniformly and so the resistance per unit length for the entire system can be computed by

$$R_{\text{tot}} = R_\square \left(\frac{1}{W} + \frac{1}{W_{gnd}}\right) \qquad (5.49)$$

At higher frequencies, though, currents redistribute to minimize the energy of the system as shown in part (b) of Fig. 5.3. The magnetic energy of the system can be minimized by minimizing the inductance value per unit length. Thus, the ground currents redistribute and concentrate under the signal carrying conductor.[2] The ground current will flow as close as possible to the signal carrying current in order to minimize the magnetic field (since ground currents flow in the opposite direction). The high frequency resistance is thus closer to

$$R_{\text{tot}} \approx 2\frac{R_\square}{W} \qquad (5.50)$$

Another perspective is to note that AC current takes the path of least impedance. Now consider the inductance value of the various paths through the ground plane. As shown in Fig. 5.4 a path far removed from the substrate has a large

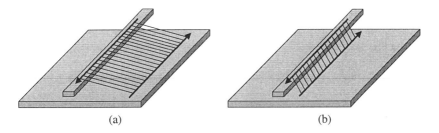

Figure 5.4. The cross-sectional area of a return path loop (a) far removed from the conductor and (b) near the conductor.

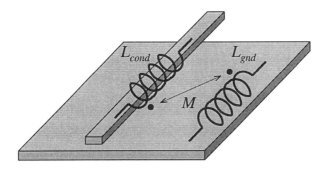

Figure 5.5. The self and mutual inductance of the conductor and return path.

cross-sectional area and thus a large value of inductance. A path directly under the inductor involves the smallest area and thus the smaller value of inductance. Therefore, AC current will flow in paths of low inductance.

A final perspective to this problem can be drawn from our study of "partial" inductance. As shown in Fig. 5.5, the total inductance for a typical path in the ground plane is given by

$$L_{\text{tot}} = L_{cond} + L_{gnd} - 2M \tag{5.51}$$

Since M varies from path to path whereas the L for a typical path is constant, nature chooses a path to maximize M so as to minimize the total inductance.

It is interesting to note that the high frequency current distribution is a result of the electric field term of

$$\mathbf{F} = q(\mathbf{E} + \mathbf{v} \times \mathbf{B}) \tag{5.52}$$

rather than the magnetic field term. This is a result of the electromagnetic interaction or Faraday's law. To see this is true consider the static magnetic force on two current carrying wires. It is well known that if the currents are in opposite directions, the wires repel one another. So based on the magnetic

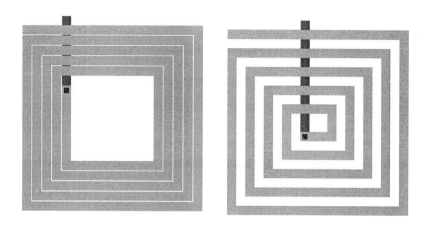

Figure 5.6. Two inductors realized with different metal spacing values of 1 μm and 10 μm.

forces alone the current distribution in our example should be the exact opposite of Fig. 5.3b.

6.2 EXAMPLE CALCULATION

We will now illustrate the importance of calculating the partial inductance matrix at high frequency with an example. Consider the two spiral inductor layouts shown in Fig. 5.6. The inductor geometries are identical with outer length $L = 200$ μm, and five turns of metal pitch $W = 10$ μm. In (a), though, the metal spacing $S = 1$ μm whereas in (b) $S = 10$ μm.

We analyze both devices by computing the partial inductance matrix $\tilde{Z}$. Each segment of the spirals is sub-sectioned into twenty separate conductors producing a 400×400 matrix $\tilde{Z}$. The matrix is reduced in order by computing $Z^M = (S\tilde{Z}^{-1}S^T)^{-1}$ and then by summing over the elements to obtain

$$R(f) + j\omega L(f) = \sum_{i,j} Z^M_{i,j}(f) \qquad (5.53)$$

From an inductance point of view, using the smaller value of spacing results in a higher DC inductance value of 5.2 nH as opposed to 2.6 nH. From Fig. 5.7 we see that as a function of frequency, the inductance values of both devices decrease slightly approaching the external inductance limit[3], falling about 4% from the DC value at 5 GHz.

The series resistance of the device, though, increases much more rapidly for the smaller value of spacing, as shown in Fig. 5.7b, where the ratio of AC to DC resistance is plotted. While the device with large spacing shows a 40% increase in series resistance at 5 GHz compared to DC, the device with small spacing suffers nearly twice as much loss at 5 GHz compared to low frequency. This

Inductance Calculations 71

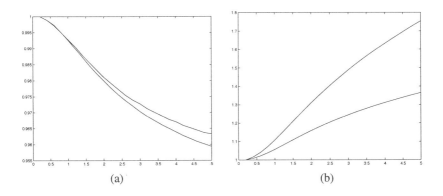

Figure 5.7. The ratio of AC to DC inductance and resistance as a function of frequency plotted for two inductors of Fig. 5.6.

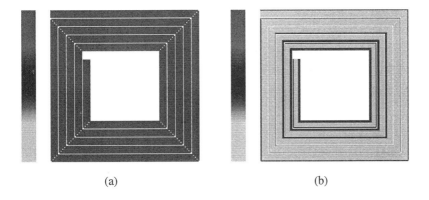

Figure 5.8. The current density at (a) 1 GHz and (b) 5 GHz.

Figure 5.9. The current density at (a) 1 GHz and (b) 5 GHz.

increase is faster than the $\sqrt{f}$ increase predicted by simple skin effect theory. To see why, note that the magnetic field is non-uniform across the cross-section of the inductor and increases rapidly towards the center of the device. Thus the magnetic field penetrating the conductors results in eddy currents with a more profound influence over the loss at the inner core of the spiral. A plot of the current density[4], shown in Fig. 5.8, shows that the 1 μm design suffers from current constriction at 5 GHz in the innermost turn. On the other contrary, as shown in Fig. 5.9, the 10 μm design has uniform current distribution even at 5 GHz, more uniform than at 1 GHz.

Now consider an optimization routine searching for a value of spacing S to optimize the Q of the device. If the optimizer makes the assumption that resistance increases like $\sqrt{f}$ independent of spacing, then clearly the optimizer will select the smallest possible value of S to maximize the inductance. In other words, an optimizer will find an incorrect optimal solution, an artifact created by the weakness of the modeling process.

Notes

1. Magnetic vector potential also has physical meaning in the study of quantum mechanics [Feynman et al., 1963].
2. For simplicity we have ignored the current distribution in the signal conductor.
3. Please note that this is the physical inductance of the device and not the effective inductance. In other words, the capacitive effects have been completely ignored in computing (5.53).
4. See Chapter 7 for an explanation of how these plots were generated. Note that these plots show the current distribution after solving the electric and magnetic problem.

Chapter 6

CALCULATION OF EDDY CURRENT LOSSES

1. INTRODUCTION

Due to the non-zero resistivity of the metal layers there are ohmic losses in the metal traces as well as eddy current losses. The eddy currents in the metal traces arise from the magnetic fields generated by the device that penetrate the metal layers. These magnetic fields induce currents that give rise to a non-uniform current distribution along the width and thickness of conductors pushing current to the outer skin of the conductors. These effects are also known as skin and proximity effects. Skin effect losses are from the magnetic field of the "self" inductance of a metal trace whereas proximity effects result from the magnetic field of nearby conductors. The proximity of nearby conductors also contributes to the current distribution in a conductor, most prominently for the innermost turns of a spiral where the magnetic field is strongest [Craninckx and Steyaert, 1997, Huan-Shang et al., 1997].

In Chapter 5 a technique was presented to analyze the skin and proximity effect losses based on the previous work of [Ruehli, 1972, Weeks et al., 1979], especially the PEEC formulation [Ruehli and Heeb, 1992]. Electrical substrate losses were also analyzed in [Niknejad and Meyer, 1998] based on the work of [Niknejad et al., 1998, Gharpurey and Meyer, 1996]. Eddy current losses in the bulk Si substrate, though, have not been accounted for thus far since a free-space Green function was used to derive the inductance. In this chapter the previous work is extended by including the magnetically induced losses in the substrate.

The importance of modeling such effects was not initially realized as these effects were not widely observable in the bipolar and BiCMOS substrates of interest because of the widespread use of highly resistive bulk materials. These effects, though, were seen to be of integral importance when researchers attempted the construction of high Q inductors over an epitaxial CMOS substrate [Niknejad,]. In [Craninckx and Steyaert, 1997] the importance of modeling eddy currents was further demonstrated through numerical electromagnetic simulation. These simulations and measurement results clearly show that eddy currents are a dominant source of loss in these substrates.

In this chapter approximate 2D and 3D expressions for the eddy current losses over a multi-layer substrate are derived. These can be used to predict the losses in inductors and transformers fabricated over such substrates. The results are derived using quasi-static analysis.

2. ELECTROMAGNETIC FORMULATION
2.1 PARTIAL DIFFERENTIAL EQUATIONS FOR SCALAR AND VECTOR POTENTIAL

Consider a long filament sitting on top of a multi-layer substrate. A cross-section of the geometry is shown in Fig. 6.1. Assume the filament is carrying a time harmonic current. The substrate is assumed infinite in extent in the traverse direction whereas each substrate layer k has thickness t_k, conductivity σ_k, magnetic permeability μ_k, and electric permittivity ϵ_k. The substrate is most likely non-magnetic or weakly diamagnetic, a good approximation for Si and other semi-conductors. The introduction of a linear magnetic substrate, though, does not complicate the analysis. The filament is a distance b above the substrate, parallel to the z-direction.

The electric and magnetic fields are completely determined by Maxwell's equations. The time harmonic fields are determined by the scalar and vector potentials [Ramo et al., 1994]

$$\mathbf{E} = -j\omega\mathbf{A} - \nabla\phi \tag{6.1}$$

$$\mathbf{B} = \nabla \times \mathbf{A} \tag{6.2}$$

For obvious reasons, we will denote the first term of (6.1) the magnetic response and the second term of (6.1) the electric response. From Maxwell's equations we have the well-known relation

$$\nabla(\nabla \cdot \mathbf{A}) - \nabla^2 \mathbf{A} = \mu \mathbf{J} + j\omega\mu\epsilon\mathbf{E} \tag{6.3}$$

Assuming the substrate and metal conductors are linear and isotropic gives

$$\mathbf{J} = \sigma\mathbf{E} + \mathbf{J}_{src} \tag{6.4}$$

Substituting (6.4) and (6.1) in (6.3) and invoking a Coulomb gauge results in the following

$$\nabla^2 \mathbf{A} = \mu(j\omega\sigma\mathbf{A} - \omega^2\epsilon\mathbf{A}(\sigma + j\omega\epsilon)\nabla\phi - J_{src}) \tag{6.5}$$

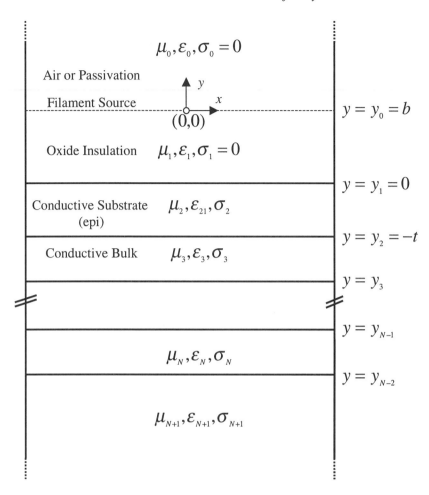

Figure 6.1. Multi-layer substrate excited by a filamental current source.

The parenthetical expression on the right hand side has units of current density. The first term can be identified as the magnetically induced eddy currents that flow in the substrate and metal conductors. The second term is the dynamic radiation current term. The third term includes the electrically induced conductive and displacement currents flowing in the substrate. Finally, the last term is the impressed currents flowing in the metal conductors.

At microwave frequencies of interest (< 15 GHz) the constant of the second term is at least three orders of magnitude smaller than the first and can be safely ignored. The physical significance is that radiation is negligible. Dropping the third term of (6.5) has two implications. First, the magnetic field contribution of the electrically induced currents will be ignored. Second, the electrically induced substrate losses will be ignored. The second implication is a far bigger concern because the electrically induced substrate losses are significant at frequencies of interest. The contribution to the magnetic field, though, can be safely ignored. To understand this physically, consider the schematic representation of the substrate currents shown in Fig. 2.4. Clearly the electrically induced current distribution leads to a zero magnetic field. This can be shown at low frequency by noting that $\nabla \times \nabla \phi \equiv 0$.

Applying the Coulomb gauge to the electric divergence relation we obtain the well-known electrostatic Poisson's equation

$$\nabla \cdot \mathbf{E} = \nabla \cdot (-j\omega \mathbf{A} - \nabla \phi) = -\nabla^2 \phi = \rho/\epsilon \qquad (6.6)$$

As discussed in Chapter 4, if we modify the above equation by replacing the electric permittivity by

$$\epsilon = \epsilon' + j\epsilon'' - j\frac{\sigma}{\omega} \qquad (6.7)$$

we account for the loss tangent of the material as well as the conductive losses. Thus the electrically induced losses can be derived from (6.6) instead of (6.5). This is valid as long as skin effect in the bulk does not significantly alter the electrically induced current distribution in the substrate. With these simplifications we have

$$\nabla^2 \phi = \frac{\rho}{\epsilon} \qquad (6.8)$$

$$(\nabla^2 - \gamma^2)\mathbf{A} = \mu \mathbf{J}_{src} \qquad (6.9)$$

where

$$\gamma^2 = \mu\epsilon\omega^2 - j\omega\sigma \qquad (6.10)$$

and ϵ is given by (6.7).

2.2 BOUNDARY VALUE PROBLEM FOR SINGLE FILAMENT

Under a two-dimensional approximation, the magnetic vector potential is directed in the direction of current, and hence has only a non-zero component in the z-direction. At microwave frequencies of interest, (6.9) simplifies and for each region

$$\nabla^2 \mathbf{A}_k = j\omega\mu_k\sigma_k \mathbf{A}_k \qquad (6.11)$$

By the method of separation of variables in rectangular coordinates [Trim, 1990], we write the solution in each layer as follows

$$A_k(x,y) = X_k(x)Y_k(y)\hat{z} \qquad (6.12)$$

Substitution of the above form in (6.11) produces two ordinary constant-coefficient second-order differential equations

$$\frac{d^2 X_k}{dx^2} = -m^2 X_k \tag{6.13}$$

$$\frac{d^2 Y_k}{dy^2} = -\gamma_k^2 Y_k \tag{6.14}$$

with the additional constraint that

$$\gamma_k^2 - m^2 = j\omega\mu_k\sigma_k \tag{6.15}$$

Due to the even symmetry of the problem one selects

$$X_k = \cos(mx) \tag{6.16}$$

and by (6.15) it follows that

$$Y_k = M_k e^{\gamma_k y} + N_k e^{-\gamma_k y} \tag{6.17}$$

Since we seek the vector potential over an infinite domain, the most general solution has the following form

$$A_k(x, y) = \int_0^\infty \left(M_k e^{\gamma_k y} + N_k e^{-\gamma_k y}\right) \cos mx \, dm \tag{6.18}$$

For N conductive layers there are $2(N + 2)$ unknown coefficients in the expansion of (6.18). There are $2(N + 1)$ boundary conditions which hold at the interface of each layers. The boundary conditions follow from Maxwell's equations [Ramo et al., 1994]

$$(\mathbf{B}_{k+1} - \mathbf{B}_k) \cdot \hat{n} = 0 \tag{6.19}$$

$$(\mathbf{H}_{k+1} - \mathbf{H}_k) \times \hat{n} = \mathbf{K} \tag{6.20}$$

where $\mathbf{K}$ is the surface current density. For $k \geq 1$ the above relations simplify to

$$B_{k,y} = B_{k+1,y} \tag{6.21}$$

$$\frac{1}{\mu_k} B_{k,x} = \frac{1}{\mu_{k+1}} B_{k+1,x} \tag{6.22}$$

where

$$\mathbf{B}_k = \nabla \times (A_k \hat{z}) \tag{6.23}$$

$$\mathbf{H}_k = \frac{1}{\mu_k} \mathbf{B}_k \tag{6.24}$$

Note that (6.21) and (6.22) must hold for each mode of (6.18), so one can show that

$$\begin{pmatrix} M_{k+1} \\ N_{k+1} \end{pmatrix} = \frac{1}{2} \begin{pmatrix} (1+\lambda_k)e^{-g_k} & (1-\lambda_k)e^{-h_k} \\ (1-\lambda_k)e^{+h_k} & (1+\lambda_k)e^{+g_k} \end{pmatrix} \begin{pmatrix} M_k \\ N_k \end{pmatrix} \quad (6.25)$$

where

$$\lambda_k = \frac{\mu_{k+1}}{\mu_k} \frac{\gamma_k}{\gamma_{k+1}} \quad (6.26)$$

and

$$g_k = (\gamma_{k+1} - \gamma_k) y_k \quad (6.27)$$
$$h_k = (\gamma_{k+1} + \gamma_k) y_k \quad (6.28)$$

Since $A \to 0$ as $y \to \pm\infty$ it follows that $M_0 \equiv 0$ and $N_{N+1} \equiv 0$ to satisfy the boundary condition at infinity.

The boundary conditions at the filament interface $y = b$ require special care. Applying (6.21,6.22) we have [Stoll, 1974]

$$\frac{1}{\mu_0} \frac{\partial A_0}{\partial x}\bigg|_{y=b} = \frac{1}{\mu_1} \frac{\partial A_1}{\partial x}\bigg|_{y=b} \quad (6.29)$$

$$\left(\frac{1}{\mu_0} \frac{\partial A_0}{\partial x} - \frac{1}{\mu_1} \frac{\partial A_1}{\partial x} \right)_{y=b} = K(x) \quad (6.30)$$

where

$$K(x) = \delta(x) I \quad (6.31)$$

or equivalently, expressing (6.31) as an inverse cosine transform

$$K(x) = \frac{I}{\pi} \int_0^\infty \cos mx \, dm \quad (6.32)$$

Thus all the unknown coefficients may be evaluated and the boundary value problem is solved. This is the approach followed by [Stoll, 1974] and [Poritsky and Jerrard, 1954]. An alternative derivation which leads to a different integral representation of the magnetic potential is presented in [Tegopoulos and Kriezis, 1985]. Observe that the magnetic field in the free-space region above the substrate may be expressed as arising from two sources, the filament current and the currents flowing in the substrate (the eddy currents). To derive the term arising from the filament in free-space observe that

$$B(r) = \frac{\mu_0 I}{2\pi r} \quad (6.33)$$

which may be expressed by the converging Fourier integrals

$$B_{0x} = \frac{\mu_0 I}{2\pi} \int e^{-|b-y|m} \cos mx \, dm \quad (6.34)$$

$$B_{0xy} = \frac{\mu_0 I}{2\pi} \int e^{-|b-y|m} \sin mx\, dm \qquad (6.35)$$

This observation implies that

$$M_0(m) = \frac{\mu_0 I}{2\pi} \frac{e^{-bm}}{m} \qquad (6.36)$$

Using the above relation and (6.29) the coefficients can be obtained uniquely for all layers. More generally, we can write

$$A(x,y) = \frac{\mu I}{2\pi} \int_0^\infty \frac{e^{m|y-y_0|}}{m}(1 + \Gamma(m))\cos(m(x-x_0))dm \qquad (6.37)$$

where (x_0, y_0) is the source filament location. The unity term accounts for the filament current in free-space and the term involving Γ accounts for the eddy currents in the substrate. In other words, the first term is the solution of the free-space problem for the impressed filamental currents whereas the second term is due to the response eddy currents in the substrate. This particular form will be very fruitful in our later analysis.

2.3 PROBLEMS INVOLVING CIRCULAR SYMMETRY

When the current excitation is circular or approximately symmetric, as in the case of a polygon spiral inductor, the assumption of circular symmetry also leads to a one-dimensional integral expression for the magnetic vector potential. The analogous solution involves Bessel functions in the place of the cosine function of equation (6.37). This problem has been treated extensively in [Tegopoulos and Kriezis, 1985, Hurley and Duffy, 1995, Hurley and Duffy, 1997] using a magnetostatic formulation and in [Wait, 1982, Mahmoud and Beyne, 1997] using an electromagnetic formulation. In this paper we will concentrate on the infinite rectangular solution as it applies more directly to devices involving orthogonal or Manhattan geometry. It should also be noted that [Lee et al., 1998a] used the circularly symmetric solution to calculate the substrate losses.

2.4 MAGNETIC VECTOR POTENTIAL IN 3D

In this section, we extend our results for eddy current losses from 2D to 3D. To do this, we first derive the lossless 3D magnetic vector potential in free-space in order to gain insight into the problem.

2.4.1 FREE-SPACE SOLUTION

In a similar vein, the magnetic vector potential may be derived directly by the governing equation (6.9) by the method of separation of variables. For a

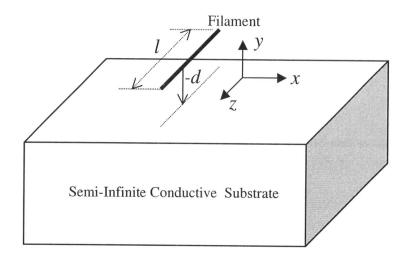

Figure 6.2. Single layer semi-infinite substrate excited by a filamental current source.

filament of length ℓ the 3D current density is given by

$$\mathbf{J}(x,y,z) = \hat{z}\delta(x)\delta(y)\left[u(z+\ell/2) - u(z-\ell/2)\right] \quad (6.38)$$

where the function $u(z)$ is the unit-step function. Enforcing the boundary conditions as before, we arrive at the following expression

$$A(x,y,z) = \int_0^\infty \int_0^\infty \frac{-\mu_0 \sin(n\ell/2)}{\pi^2 \sqrt{m^2+n^2}} 2\sinh\left(\sqrt{m^2+n^2}|y-b|\right)$$
$$\times \cos(mx)\cos(nz) dm dn \quad (6.39)$$

Note that this integral may be computed in closed-form as presented in (5.42). Our purpose in re-deriving the above expression in open-form will become clear in the next section.

2.4.2 3D SOLUTION OVER A LOSSY SUBSTRATE

Consider the geometry of the 3D problem shown in Fig. 6.2. The general solution for the magnetic vector potential for a finite filament over a lossy substrate in 3D is given by

$$A_1 = \int_0^\infty \int_0^\infty N_1 e^{-\sqrt{m^2+n^2}\,y} \cos(mx)\cos(nz) dm dn \quad (6.40)$$

above the filament and by the following expression

$$A_2 = \int_0^\infty \int_0^\infty \left(M_2 e^{\sqrt{m^2+n^2}\,y} + N_2 e^{-\sqrt{m^2+n^2}\,y}\right) \cos(mx)\cos(nz) dm dn \quad (6.41)$$

above the substrate and below the filament. Applying the boundary conditions we arrive at the following solution

$$M_2 = \frac{-\mu_0}{2\pi^2} \frac{\sin(n\ell/2)}{n\gamma_2} \tag{6.42}$$

$$N_2 = -\frac{\gamma_3 - \gamma_2}{\gamma_3 + \gamma_2} e^{2\gamma_2 d} M_2 \tag{6.43}$$

$$N_1 = \left[1 - \frac{\gamma_3 - \gamma_2}{\gamma_3 + \gamma_2} e^{2\gamma_2 d}\right] M_2 \tag{6.44}$$

where $\gamma_2 = \sqrt{m^2 + n^2}$ and $\gamma_3 = \sqrt{m^2 + n^2 + j\mu\omega\sigma}$. The expression for N_1 can be expanded into the following form

$$\frac{(j\omega\sigma\mu)N_1}{K} = \underbrace{\frac{(j\omega\sigma\mu)}{n\gamma_2}}_{\text{pure inductance}} - \underbrace{(j\omega\sigma\mu)\frac{e^{-2\gamma_2 d}}{n\gamma_2}}_{\text{image inductance}} + 2\gamma_3 e^{-2\gamma_2 d}\frac{1}{n\gamma_2}$$

$$-\frac{2\gamma_2 e^{-2\gamma_2 d}}{n} \tag{6.45}$$

where $K = n\gamma_2 M_2$.

Note the first and third term are purely imaginary and frequency-independent and therefore only represent the inductive portion, and not the loss, of the magnetic vector potential. In particular, from (6.38) we may identify these terms as arising from two equivalent finite length filaments situated in free-space carrying equal and opposite currents. From (5.42) this contribution to the integral can be factored out and computed in closed-form. This results in tremendous savings in numerical computation due to the avoidance of the logarithmic singularity. The other terms represent loss and reflected inductance

$$j\omega\sigma\mu \tilde{N}_1 = 2e^{-2\gamma_2 d}\frac{K}{n}(\gamma_3 - \gamma_2) \tag{6.46}$$

Note that $\gamma_3 = \gamma_2$ at zero frequency so this term drops out at DC. The physical interpretation is that there are no DC losses since eddy currents are electromagnetic phenomena. Simplifying the loss portion of the above integral we have

$$A_{loss}(x,y,z) = \frac{\mu_0 I}{\pi^2 j(\omega\sigma\mu_0)} \int_0^\infty \int_0^\infty dm\,dn \cos(nz)\cos(mx) \times$$
$$\left[\frac{\sin(n\ell/2)}{n}\right] e^{-\gamma_2(y+2d)}(\gamma_3 - \gamma_2) \tag{6.47}$$

If we integrate the above expression over the path of a parallel filament separated by a distance s we obtain the induced voltage due to the eddy current losses

$$V_{loss} = j\omega \int_{-\ell/2}^{\ell/2} A_{loss}(s,0,z)dz \tag{6.48}$$

84 INDUCTORS AND TRANSFORMERS FOR SI RF ICS

which yields

$$Z = \frac{V_{loss}}{I} = \frac{2}{\pi^2 \sigma} \int_0^\infty \int_0^\infty \left(\frac{\sin(n\ell/2)}{n}\right)^2 \cos(ms) e^{-\gamma_2 2d} (\gamma_3 - \gamma_2) dm dn \quad (6.49)$$

where the term $\Re[Z]$ only contains the substrate reflected losses and $\Im[Z]$ represents the substrate reflected inductance.

3. EDDY CURRENT LOSSES AT LOW FREQUENCY
3.1 EDDY CURRENT LOSSES FOR FILAMENTS

With the magnetic vector potential known, we can proceed to calculate the eddy current losses. There are two approaches to determining the losses. One approach is to use Poynting's theorem to calculate the total power crossing a surface enclosing the filament. In the time harmonic case the real component of this power must be due to the lossy substrate since no other loss mechanisms are present [Tegopoulos and Kriezis, 1985]. The complex Poynting's vector is given by

$$\mathbf{S} = \frac{1}{2}(\mathbf{E} \times \mathbf{H}^*) \quad (6.50)$$

If we integrate the normal component of this vector over the surface $y = 0$ we obtain the power crossing the substrate

$$P + jQ = \frac{1}{2} \int_{-\infty}^{\infty} (\mathbf{E} \times \mathbf{H}^*) \cdot \hat{y} dx \quad (6.51)$$

Considering now only the magnetic response of the substrate, from (6.1) we have

$$\mathbf{E} = -j\omega \mathbf{A} \quad (6.52)$$

Thus, (6.50) becomes

$$\mathbf{S} = \frac{-j\omega}{2\mu} \mathbf{A} \times \nabla \times \mathbf{A} \quad (6.53)$$

For the geometry of Fig. 6.1 the integrand of (6.51) simplifies to

$$(\mathbf{E} \times \mathbf{H}^*) \cdot \hat{y} = \frac{-j\omega}{\mu} A \frac{\partial \mathbf{A}^*}{\partial y} \quad (6.54)$$

In section 2.2 it was shown that the magnetic vector potential has the following general form

$$A(x,y) = \frac{\mu I}{2\pi} \int_0^\infty f(y,m) \cos mx \, dm \quad (6.55)$$

Differentiating (6.55) under the integral and substituting in (6.51) results in

$$P + jQ = \frac{-j\omega}{2\mu} \int_{-\infty}^{\infty} dx \times$$

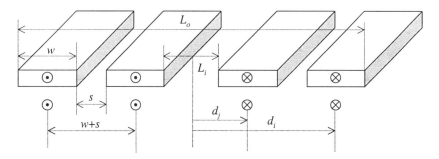

Figure 6.3. Cross-section of square spiral inductor.

$$\left[\frac{\mu I}{2\pi}\int_0^\infty \frac{\partial f(y,n)^*}{\partial y}\cos nx\,dn\right]\left[\frac{\mu I}{2\pi}\int_0^\infty f(y,m)\cos mx\,dm\right] \quad (6.56)$$

If we interchange the order of integration and observe that

$$\lim_{L\to\infty}\int_{-L}^{L}\cos mx\cos nx\,dx = \lim_{L\to\infty}\left[\frac{\sin L(m-n)}{m-n} + \frac{\sin L(m+n)}{m+n}\right]$$
$$= \pi(\delta(m-n)+\delta(m+n)) \quad (6.57)$$

we obtain

$$P + jQ = \frac{-j\omega\mu I^2}{8\pi}\int_0^\infty f(y,m)f_y^*(y,m)\,dm \quad (6.58)$$

Thus, the equivalent resistance per unit length seen by the source driving the filament becomes

$$R_{eq} = \Re[2(P+jQ)] \quad (6.59)$$

The imaginary part of (6.58) also contains useful information as it represents the reactive power crossing the surface which can be attributed to inductance. This is a negative increasing function of frequency which represents decreasing inductance as a function of frequency. The inductance decreases due to the "image" eddy currents flowing in the substrate. By Lenz's law, these currents flow in a direction opposite to the impressed current and hence generate a magnetic field that tends to cancel the penetrating magnetic field of the source, thereby decreasing the inductance.

Using this principle let us derive the power loss for the configuration shown in 6.3. Note that two sets of N parallel current filaments carry a current I where the individual filaments are separated by a distance s and the two sets of filaments are separated by a distance d. Notice that this current distribution crudely approximates half of the current distribution for a spiral inductor of N turns. In a spiral inductor the filaments have finite length and vary in length.

Here we neglect "end-effects" and calculate the losses for the average length filament.

Using (6.55) we have

$$A(x,y) = \frac{\mu_0 I}{2\pi} \int_0^\infty f(y,m) \left(\sum_{i=1}^N \cos m(x - d_i) - \cos m(x + d_i) \right) dm \quad (6.60)$$

and

$$A_y^*(x,y) = \frac{\mu_0 I}{2\pi} \int_0^\infty f_y^*(y,n) \left(\sum_{i=1}^N \cos n(x - d_i) - \cos n(x + d_i) \right) dn \quad (6.61)$$

and applying (6.58) while changing the order of integration we have

$$P + jQ = \frac{\mu_0^2 I^2}{4\pi^2} \int_0^\infty \int_0^\infty f_y^*(m) f(n) \left(\int_{-\infty}^\infty H(x) dx \right) dm dn \quad (6.62)$$

where

$$H(x) = \sum_{i,j} \pm \cos\left(n(x \pm d_i)\right) \cos\left(m(x \pm d_j)\right) \quad (6.63)$$

where $H(x)$ has been written in shorthand notation. Each x domain integral of (6.62) takes the form of

$$\int_{-\infty}^\infty \cos m(x + \alpha) \cos n(x + \beta) dx$$
$$= \delta(m - n) \cos(m\alpha + n\beta)\pi + \delta(m + n) \cos(m\alpha - n\beta)\pi \quad (6.64)$$

Using the above relation reduces (6.62) to

$$P + jQ = \frac{\mu_0^2 I^2}{\pi} \int_0^\infty f_y^*(m) f(n) \left(\sum_{i,j} \sin(m d_j) \sin(m d_i) \right) dm \quad (6.65)$$

Alternatively, one can derive the equivalent impedance per unit length seen by the source driving the filament by simply observing that by (6.1) the reflected magnetic contribution to the impedance must be [Stoll, 1974]

$$R_{eq} = \Re \left[\frac{j\omega A_0(0,b)}{I} \right] \quad (6.66)$$

Notice that (6.66) will lead to a different yet equivalent integral expression for the eddy current losses.

3.2 EDDY CURRENT LOSSES FOR CONDUCTORS

Due to the linearity of Maxwell's equations, we can invoke the superposition principle to calculate the losses when more than one filament is present, even for a continuous distribution of the field. An alternative viewpoint is that in calculating the vector potential for the filament case we have actually derived the kernel of the integral operator that is the inverse transform of (6.11), or the Green function [Roach, 1982].

Thus, for any distribution of current over the multi-layer substrate of 6.1 we can write the resulting vector potential as [1]

$$A(x,y) = \int\int G(x,y) J(x,y) dS \quad (6.67)$$

where the surface integral is taken over the cross-section of the conductor and has the form of (6.55). If the current distribution is uniform this simplifies to

$$A(x,y) = I \int\int G(x,y) dS \quad (6.68)$$

In many practical cases the current distribution is non-uniform. In these cases one may approximate the current distribution by dividing the cross-section into uniform current distribution segments and apply (6.68) to such segments. This is discussed in more detail in [Weeks et al., 1979, Ruehli and Heeb, 1992, Kamon et al., 1994a, Niknejad and Meyer, 1998].

Integrating (6.55) over the width w of the source conductor we obtain

$$A_w(x,y) = \frac{\mu I}{2\pi} \int_0^\infty f(y,m) \left[\frac{\sin\frac{mw}{2}}{\frac{mw}{2}} \right] \cos mx\, dm \quad (6.69)$$

If we further average the above expression over the finite width of the field point we obtain

$$A_{ww}(x,y) = \frac{\mu I}{2\pi} \int_0^\infty f(y,m) \left[\frac{\sin\frac{mw}{2}}{\frac{mw}{2}} \right]^2 \cos mx\, dm \quad (6.70)$$

assuming the field conductor width is also equal to w.

In order to calculate the total impedance for a set of filaments in series, one must account for the self and mutual impedance terms

$$Z_{eq} = \sum_{i,j} \frac{j\omega A_j(d_i,b)}{I} = \sum_{i,j} Z_{ij} \eta_j \quad (6.71)$$

where $A_j(d_i,b)$ is the vector potential generated by the jth conductor evaluated at the location of conductor i and is given by

$$A_j(d_i,b) = \pm \frac{\mu I}{2\pi} \int_0^\infty f(b,m) \cos m(d_i - d_j)\, dm \quad (6.72)$$

where the positive sign is used when the currents flow in the same direction whereas the negative sign is used when the currents flow in opposite directions.

The factor $\eta_i = I_i/I$ accounts for the non-uniform current distribution along the length of the device. At low frequencies, $\eta_i \approx 1 \, \forall \, i$, since no current is lost to the substrate due to displacement current. At higher frequencies, though, it is critical to evaluate (6.71) with this factor in place as the current distribution becomes non-uniform. In the next section we derive this current distribution.

4. EDDY CURRENTS AT HIGH FREQUENCY
4.1 ASSUMPTIONS

In [Ruehli and Heeb, 1992] the PEEC formulation is shown to be equivalent to solving Maxwell's equations. We can thus formulate our problem using a modified PEEC technique. Our modifications mainly take advantage of the special geometry and symmetries in the problem to reduce the calculations. This approach has already been pursued in [Niknejad and Meyer, 1998]. Here, we present a more symmetric formulation.

First, we would like to avoid generating volume elements in the substrate. That would allow free-space Green functions to be employed but would produce too many elements. Since the Si substrate is only moderately conductive, we would require several skin depths of thickness in the substrate volume as well as an area at least 2-3 times the area of the device under investigation to include the fringing fields. Since the fields would vary rapidly across the cross-sectional area and depth of the substrate, many mesh points would be required. On the other hand, if we formulate the problem with a multi-layer Green function, the substrate effects are taken care of automatically and the substrate can be effectively ignored in the calculation. Therefore, only the conductor volumes need to be meshed.

Furthermore, since the conductors that make up the device are good conductors, consisting typically of aluminum, gold, or copper, displacement current in the volume of the conductors can be safely ignored. Thus, the divergenceless current distribution in the conductors is found solely by solving the magnetostatic problem (6.9). The divergence of the current, or charge, is determined solely from the electrostatic distribution of charge found by solving (6.8).

One further assumption greatly reduces the order of the problem. If we assume that the current flows along the length of the conductors in a one-dimensional fashion, then only meshing in one dimension as opposed to two or three dimensions is needed. For the typical square spiral shown in Fig. 2.8, we see that this is indeed a good approximation. Note that this does not preclude a non-uniform current distribution along the length, width, or thickness of the conductors. Rather, the current is constrained to flow in one direction only. This assumption is mostly in error around the corners of the device where we

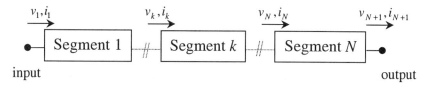

Figure 6.4. Voltages and currents along series-connected 2-port elements.

may choose to use a two-dimensional current distribution or we may simply ignore the corner contributions.

4.2 INDUCTANCE MATRIX

Given the assumptions of section 4.1, we may sub-divide the device into many sub-conductors as shown in 6.4. Since the current is constrained to flow in one dimension, the problem can be reduced by solving the equivalent magnetic circuit equations. For the system of filaments we calculate a partial inductance matrix $\tilde{Z}^M$ [Ruehli, 1972] where each non-diagonal element is computed with

$$\tilde{Z}^M_{i,j} = j\omega \int_{C_i} A_j \cdot dl_i \tag{6.73}$$

and the diagonal elements are given by

$$\tilde{Z}^M_{i,i} = R_i + j\omega \int A_i \cdot dl_i \tag{6.74}$$

Employing the same approximations as [Weeks et al., 1979] we reduce this matrix to the level of the conductors by invoking KCL at each node to obtain [Kamon et al., 1994a]

$$Z^M = \left(S^T \left(\tilde{Z}^M \right)^{-1} S \right)^{-1} \tag{6.75}$$

where the sparse rectangular matrix S sums over the current sub-elements of a conductor. Thus, each row has a one in a position corresponding to a sub-element and zero otherwise. The problem with computing (6.75) directly is that the large matrix $\tilde{Z}^M$ must be computed and inverted.

4.3 FAST COMPUTATION OF INDUCTANCE MATRIX

In [Kamon et al., 1994a] computation of (6.75) is avoided altogether by an iterative solution. The matrix-vector products are accelerated by taking advantage of the $1/R$ form of the free-space kernel [Greengard and Rokhlin, 1997]. This kernel specialization, though, limits the applicability of the technique and precludes its application to the problem at hand since this would require us to either ignore the Si substrate (which distorts the free-space Green function) or to mesh the substrate. Not only does the substrate meshing unnecessarily increase the size of the problem, but it also requires a more complete PEEC formulation since displacement current cannot be ignored in the substrate.

The authors of [Kapur and Long, 1997] have developed a more general iterative solver that can be applied to (6.75). The basis of their technique is to factor $\tilde{Z}^M$ using the singular value decomposition (SVD). Using the SVD one can compress the matrix by only retaining the larger singular values. This also allows fast computation of matrix-vector products. This, of course, requires an efficient procedure to compute the SVD. For matrices generated from integral equations, [Kapur and Long, 1997] develops an efficient recursive process to compute the SVD.

In [Niknejad and Meyer, 1998] an approximate technique is presented to compute (6.75) indirectly by ignoring long range interactions. This is in fact the crux of all the abovementioned techniques.

4.4 EFFICIENT CALCULATION OF EDDY CURRENT LOSSES

As it stands, the derivations of section 2. are not directly applicable to the above analysis unless an unrealistic two-dimensional approximation is used. A three-dimensional approach, on the other hand, requires numerical integration calculations that are at least four orders of magnitude more expensive to perform. To see this, note that instead of a one-dimensional integral for the magnetic vector potential we would require a two-dimensional integral. Also, integration of A over the source and field cross-sections will add two to four more dimensions. Finally, integration of A along the path of the field will involve one final line integral, adding at least one dimension to the problem. The two-dimensional approximation, though, only involves an integral of one dimension. This is because the integrations over the cross-sections can be performed analytically and the integration along the path of the field is trivial to compute due to the z-direction invariance inherent in the two-dimensional approach.

On the other hand, the free-space calculation of the magnetic vector potential is exact and the mutual inductance between filaments may be performed in closed form. To include the cross-section of the conductors requires numerical integration over the volume of the conductors. The geometric mean distance (GMD) approximation [Greenhouse, 1974][Grover, 1946], on the other hand, yields closed-form results for the case of parallel rectangular cross-sections. Thus, each matrix element computation can be performed in closed form. It has been found experimentally that the GMD approximation computes the free-space inductance value to a high precision for conductors over insulating or semi-insulating substrates [Krafcsik and Dawson, 1986],[Pettenpaul and et al.,

1988],[Nguyen and Meyer, 1990],[Long and Copeland, 1997],[Niknejad and Meyer, 1998].

In order to retain the accuracy of the free-space GMD approximation and the simplicity of the two-dimensional approximation, we propose a hybrid calculation. As already noted, due to linearity of the partial differential equation (6.9) we can write the general solution as follows

$$A(x,y,z) = A_{\text{free-space}} + A_{\text{substrate}} \qquad (6.76)$$

The first term is the magnetic vector potential computed in free space. The second term is the magnetic vector potential resulting from the substrate currents. Note that the substrate currents are response currents whereas the free-space currents are impressed currents. The response currents are not known *a priori* so the second term cannot be computed directly. However, we have already factored $A(x,y)$ in this form in (6.37). Thus we may compute the first term directly, using the GMD approximation to simplify the calculations. The second term is computed using the two-dimensional approximation developed in section 2.. Since the substrate effects are secondary in nature at frequencies of interest, the error in the above approximation tends to be second order yielding accurate overall results.

Hence, computation of (6.75) proceeds in two stages

$$\tilde{Z}_{i,j}^{M} = \tilde{Z}_{i,j}^{M,F} + \tilde{Z}_{i,j}^{M,S} \qquad (6.77)$$

where the second term is computed from

$$\tilde{Z}_{i,j}^{M,S} = \frac{-\mu j \omega}{2\pi} \int_0^\infty \frac{e^{-m|y-y_0|}}{m} \Gamma(m) K(m,w) \cos\left(m(x-x_0)\right) dm \qquad (6.78)$$

The real part of the above matrix element represents the eddy current losses and the imaginary part represents the decrease in inductance due to image currents flowing in the substrate. Note that the kernel K is computed by integrating over the cross-section of the source and field points. This term is unity for filaments, and for thin conductors of width w, it is given by the bracketed expression of (6.70).

Note that the purpose of calculating $\tilde{Z}^M$ is to obtain and account for the non-uniform current distribution in the volume of the conductors. This non-uniformity arises primarily due to the non-uniform mutual inductive effects that are contained in the first term of (6.77). Since the losses computed from (6.78) tend to be uniform and do not influence the skin and proximity effects, one can reduce the number of calculations of (6.78) by including the substrate reflection terms at the conductor stage rather than at the sub-conductor stage. Thus, we may include the computation of (6.78) by simply adding it to the matrix term directly. This reduces the number of computations from $N^2 \cdot M^2$ to N^2 where there are N conductors divided into an average of M sub-conductors. The validity of this approach can be verified by calculating the equivalent resistance and inductance of a device both ways.

4.5 INDUCTANCE MATRIX EDDY CURRENT LOSS FOR SQUARE SPIRAL INDUCTOR

To compute (6.78) for the case of a spiral inductor, one can take advantage of the two-dimensional symmetries of Fig. 6.3 to further reduce the number of calculations from $O(N^2)$ to $O(2N)$. (d_{src}, h_{src}) and $r_{fld}(d_{fld}, h_{fld})$ represent the (x,y) coordinates of the source and field. Also define $L_{i,j}^{GML} = \sqrt{L_i L_j}$ and

$$f_{r,i}(|d_{src} - d_{fld}|, h_{src} + h_{fld}, w) = \int_0^\infty \frac{e^{-m(h_{src}+h_{fld})}}{m} \Gamma_{r,j}(m) K(m,w) \cos\left(m\left(d_{src} - d_{fld}\right)\right) dm$$

```
Compute Z^{M,S} : 2N × 2N
begin:
Compute diagonal terms:  Z_{i,j}^{GML} = L_{i,j}^{GML} f(0, 2b, w)
for j = 2:N
        Z_{1,j}^{M,S} = L_{1,j}^{GML} f((j-1)s, 2b, w)
end
for i = 2:N
        for j = i+1:N
                Z_{i,j}^{M,S} = \frac{L_{i,j}^{GML}}{L_{i-1,j-1}^{GML}} Z_{i-1,j-1}^{M,S}
        end
end
for j = N+2:2N
        Z_{1,j}^{M,S} = L_{1,j}^{GML} f(L_i + (j-(N+2))s, 2b, w)
end
for i = 2:N
        for j = N+2:2N
                Z_{i,j}^{M,S} = \frac{L_{i,j}^{GML}}{L_{i-1,j-1}^{GML}} Z_{i-1,j-1}^{M,S}
        end
end
for i = N+1:2N
        for j = N+2:2N
                Z_{i,j}^{M,S} = \frac{L_{i,j}^{GML}}{L_{i-N,j-N}^{GML}} Z_{i-N,j-N}^{M,S}
        end
end
let Z_{j,i} = Z_{i,j}
```

The complete substrate reflection matrix may be computed using the above algorithm. Note that the above algorithm involves only $O(2N)$ computations since the double loops only involve data transfer.

5. EXAMPLES
5.1 SINGLE LAYER SUBSTRATE

The magnetostatic problem of a one layer conductive substrate has been the subject of detailed investigations. [Stoll, 1974, Poritsky and Jerrard, 1954, Tegopoulos and Kriezis, 1985] derive and compute the integrals of section 2.. In particular, [Poritsky and Jerrard, 1954] discusses numerical and analytical techniques to compute the integral. In our work we found numerical integration sufficient and so analytical integration was not our main focus. The solution of the one layer problem is summarized by the following reflection coefficient

$$\Gamma(m) = \frac{\gamma - m}{\gamma + m} e^{-2my_0} \qquad (6.79)$$

where y_0 is the source y-coordinate. For the case of a one layer substrate, we found the following analytical representation

$$Z_{ij} = -200\pi j\omega \left(g(z_1) + g(z_2) + \frac{1}{z_1^2} + \frac{1}{z_2^2} \right) \qquad (6.80)$$

where

$$z_{1,2} = (2b \pm (d_i - d_j)j)(j-1)\frac{\sqrt{800\pi\sigma\omega}}{2} \qquad (6.81)$$

and

$$g(z) = \int_0^\infty e^{-cz}\sqrt{c^2 - 1}\,dc \qquad (6.82)$$

The above integral can be represented as follows

$$g(z) = \frac{K_1(z)}{z} + \pi j\frac{I_1(z)}{2z} - \frac{jz}{3}\,_pF_q(1, \{\frac{3}{2}, \frac{5}{2}\}; \frac{z^2}{4}) \qquad (6.83)$$

where I_1 and K_1 are first-order modified Bessel functions of the first and second kind respectively, and $_pF_q$ is a generalized Hypergeometric function [Abramowitz and Stegun, 1972]. Since (6.82) represents the contour integration of an analytic function, its value is path-independent. Using this property, integral representations of the various standard mathematical functions can be used to derive the above result.

But, as previously noted, numerical integration is often faster than the direction computation of (6.83) and this approach will be pursued for the more complicated geometries where analytical results are more difficult to obtain.

5.2 TWO LAYER SUBSTRATE

For the two layer problem, the equations of section 2. are set up and involve six equations in six unknowns. The solution can be simplified into the following form

$$\Gamma(m) = e^{-2my_0} \frac{\gamma_2(m-\gamma_3) - (\gamma_2^2 - m\gamma_3)\tanh(t\gamma_2)}{\gamma_2(m+\gamma_3) + (\gamma_2^2 + m\gamma_3)\tanh(t\gamma_2)} \quad (6.84)$$

where y_0 denotes the source y-coordinate and t is the thickness of the top substrate layer. Note that (6.84) reduces to (6.79) as $t \to \infty$. It can also be shown that

$$\lim_{m \to 0} \frac{\Im(\Gamma(m))}{m} < \infty \quad (6.85)$$

Also, since (6.84) is exponentially decreasing for large m, numerical integration of (6.78) converges rapidly.

The above result along with (6.78) can be used to solve for the eddy current losses and decrease in inductance due to the conductive substrate.

Notes

1 Note that this is not in general true for the vector potential since a dyadic Green function must be employed. However, it is valid for the two-dimensional quasi-static case under investigation.

Chapter 7

ASITIC

1. INTRODUCTION

ASITIC ("Analysis and Simulation of Inductors and Transformers for Integrated Circuits") has been a major practical component of this research. *ASITIC* is the amalgamation of the key concepts and techniques described in this book, assembled into a user-friendly and efficient software tool. As illustrated in Fig. 7.1, *ASITIC* allows one to move easily between the electrical, physical, geometric, and network domains. In the electric domain, the device is described by the relevant electrical parameters, such as inductance, capacitance, quality-factor Q, and self-resonant frequency. In the physical domain, the device is described by the constituent material properties, such as the thickness, conductivity, permittivity, and permeability. In the geometric domain, the device is described by its physical dimensions and relative position in the volume of the integrated circuit. In the network domain, the device is described by network two-port parameters.

The ability to move easily from one domain to the other is an important property of *ASITIC*, allowing circuit designers and process engineers to optimize the device structure and the process for maximizing the quality of passive devices. This requires *ASITIC* to be not only an accurate tool over the frequency range of interest, but also an efficient tool. After all, highly sophisticated numerical tools such as EM solvers already exist. But such tools are comparatively slower than *ASITIC*–at least one or two orders of magnitude slower–since they solve more general problems.

ASITIC has also been designed to be a fairly flexible tool. As mentioned before, the modern IC process allows highly complicated geometrical structures to be designed over the Si substrate. The MEMs revolution is continuously expanding the possibilities as more and more complex electromechanical

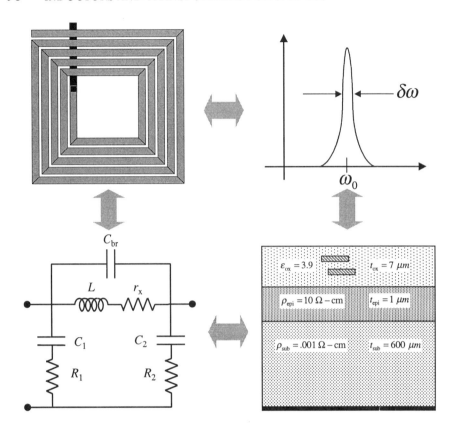

Figure 7.1. ASITIC users can move between geometric (layout), electrical (inductance, Q, self-resonance), physical (technology file), and network (two-port parameters) domains.

structures are fabricated on Si. Thus one of the major goals of *ASITIC* from the outset was to allow the analysis the analysis of an arbitrary interconnection of metal structures over the Si substrate.

In summary, the goal of *ASITIC* has been to create an easy-to-use numerical software package for the analysis and design of passive devices over the Si substrate. The key criteria for the project have been accuracy, flexibility, and efficiency.

Since 1995, *ASITIC* has been a freely available software package distributed throughout the IC and EM community by means of the Internet. To date, over 1500 universities, organizations, and commercial entities have used *ASITIC* to solve practical and experimental problems. This has resulted in great interaction between the users and creators of *ASITIC* which has fueled the continuous evolution of *ASITIC*.

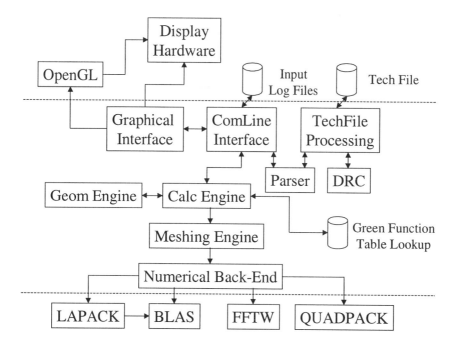

Figure 7.2. A block diagram of the *ASITIC* modules.

2. ASITIC ORGANIZATION

Fig. 7.2 is a block diagram of *ASITIC*. *ASITIC* is composed of several software modules that interact over clearly defined interfaces. The user interacts with *ASITIC* at the top level through the graphical and text interface. The technology file describes the pertinent process parameters such as substrate layer thickness, conductivity, and permittivity data, as well as metal thickness and conductivity values. By means of *ASITIC* commands, users are able to create, modify, optimize, and analyze passive devices.

The top *ASITIC* layers rely on the geometry and calculation engines to create and analyze structures. The geometry engine is able to synthesize structures such as square and polygon spirals and the calculation engine is able to quickly analyze the structures and display electrical parameters. The calculation engine in turn depends on the meshing engine to convert geometric representations of devices into electrically small geometric sub-elements used for the analysis.

The numerical back-end modules convert the electrical sub-elements into algebraic equations through numerical integration. These numerical computations are accelerated by several software libraries such as Basic Linear Algebra Subroutines (BLAS), Linear Algebra Package (LAPACK) [Dongarra and Demmel, 1991], an extension of (LINPACK), Fastest FFT in the West (FFTW)

[Frigo and Johnson, 1998], and the numerical integration package QUADPACK [et al., 1983].

Another important element in *ASITIC* has been the graphical interface. *ASITIC* is capable of displaying devices in two and three dimensions. The three-dimensional representations produced with OpenGL are highly useful in understanding and verifying complex multi-metal structures. Physical dimensions can be distorted to more easily visualize the structure. For instance, the z-direction can be scaled to clearly delineate closely spaced metal layers. *ASITIC* can also display the current and charge density in a spiral, as shown in Figures 5.8 and 5.9. This is an especially important visualization capability as it allows the device designer to understand the current flow and distribution, and hence the losses, in a device. For instance, it is simple to understand why a tapered spiral works when one can see the current constriction in the inner turns.

3. NUMERICAL CALCULATIONS

ASITIC converts Maxwell's equations into a linear system of equations with the aid of the semi-analytical Green functions. These equations are numerically stable with typical matrix condition numbers of 10, and can be solved numerically using Gaussian elimination. Since a typical device involves hundreds or at most thousands of elements, numerical packages such LAPACK can be used to efficiently invert the linear systems. LAPACK uses BLAS level 3 routines which utilize the system cache to maximize memory throughput. For larger systems, iterative solutions are more appropriate.

The transition from a geometric description of a device to the electrical properties at a given frequency involves three general steps. First, inductance and capacitance matrices must be constructed. This is done in the "matrix-fill" stage, where matrix elements are computed from numerical volume/surface integration of the underlying Green function. In the next stage, the capacitance and inductance matrix are assembled into a large system of equations by the PEEC formulation, corresponding to invoking KCL, KVL, and charge conservation, the electrical analogs of Maxwell's equations. These steps will be described in section 4.. Finally, the system of equations is solved for the electrical properties of the system.

The capacitance "matrix-fill" stage involves numerical integration of the Green function over volume elements. The quasi-static electrical computation can be performed in closed-form as shown in [Niknejad et al., 1998, Gharpurey and Meyer, 1996]. This is because the underlying Green function is described semi-numerically as a double infinite summation. Upon integration and truncation of the series, we can reduce the volume integral computation to the sum of 64 complex additions, where each complex addition is computed as an entry in the two-dimensional discrete cosine transform (DCT) of the Green function.

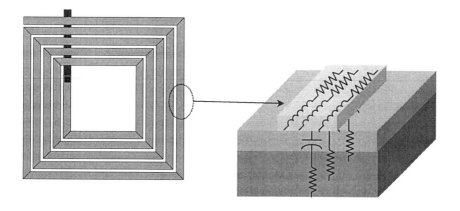

Figure 7.3. An electrically short segment of the device.

This results in tremendous savings in computation as the "matrix-fill" operation is reduced to constant time complex addition and table look-up operations.

The inductance "matrix-fill" can also be performed in closed form and is described in detail in Chapter 6. Note that (6.78) can be computed numerically using QUADPACK. This package contains code to efficiently calculate (6.78) and the results converge much faster than using Romberg integration, as is done by [Wolfram, 1999].

4. CIRCUIT ANALYSIS
4.1 MODIFIED PEEC FORMULATION

In this section we will use a modified PEEC [Ruehli, 1974] formulation to obtain 2-port parameters of various passive devices constructed over the Si substrate. This formulation is an extension of the work first presented in [Niknejad and Meyer, 1998].

Consider an arbitrary interconnection of conductors. As an example, consider a square spiral inductor as shown in Fig. 2.8. We first break this spiral inductor up into N electrically short segments, as shown in Fig. 7.3. The maximum length of each segment is given by $l_{max} = \alpha\lambda$ where λ is the quasi-TEM mode wavelength at the frequency of interest and α is a small number.

Next, we calculate the partial inductance and capacitance matrices for the system of conductors. To calculate the capacitance matrix, we further break up the N segments into panels of constant charge. Each segment is sub-divided by its width and thickness if it exceeds $t_{max} = \beta\delta$ where $\beta < 1$ and δ is the skin depth. The complex impedance matrix is computed for this system of conductors using the Green function presented in [Niknejad et al., 1998]. The real part of this matrix represents electrically induced substrate losses and the imaginary part represents the capacitive coupling through the air, oxide, and substrate.

A similar procedure is followed for the partial inductance matrix with one exception: The segments are partitioned into two groups. The first group represents segments where the direction of the current is known approximately, such as in spiral segments; the second group represents segments where the direction of the current is not known *a priori*, such as in a nearby grounded conductor. This division is important since the first group is by far the larger and knowledge of the current direction simplifies the calculations and reduces the PEEC formulation to one dimension. On the other hand, a few segments, such as corners and ground rings, will have current flowing in non-predictable ways, especially at high frequency, and a two-dimensional PEEC formulation is used in two orthogonal directions to obtain the current directions.

The capacitance and inductance matrices are next "compressed" by invoking KCL and charge conservation. This step reduces the size of the matrices to $N \times N$. The inductance matrix is now complex, with real components representing eddy currents in the conductors and in the substrate and imaginary components representing self and mutual inductance value in addition to reflected inductance from the substrate eddy currents. Note that the magnetic coupling from the electrically induced currents to the inductance matrix is ignored, a step we have justified in Chapter 4.

Now the topology of the segments is taken into account, and invoking KCL and KVL yields two-port relations.

4.2 SERIES CONNECTED TWO-PORT ELEMENTS

Consider the series interconnection of conductors as shown in Fig. 6.4. Using the reduced partial inductance matrix along with the reduced lossy capacitance matrix, one can form the following system of linear equations.

Let $i_{s,k} = \frac{1}{2}(i_k + i_{k+1})$ represent the average current flowing in the kth conductor. Similarly, let $v_{s,k} = \frac{1}{2}(v_k + v_{k+1})$ represent the average voltage of each conductor. Applying KCL and KVL at each node gives

$$i_k - i_{k+1} = \sum_{j=1}^{N} Y_{k,j}^C \left(\frac{v_j + v_{j+1}}{2} \right) \tag{7.1}$$

$$v_k - v_{k+1} = \sum_{j=1}^{N} Z_{k,j}^M \left(\frac{i_j + i_{j+1}}{2} \right) \tag{7.2}$$

where Y^C represents the complex lossy capacitive admittance matrix which is computed directly in [Niknejad et al., 1998]. Z^M is computed using (6.77). Note that these matrices are compressed or reduced in order and contain the effects of non-uniform charge and current distribution in each conductor. All loss mechanisms relevant at microwave frequencies are thus contained in these matrices. Y^C is frequency-dependent and includes electrically induced substrate losses. Z^M includes ohmic losses, skin and proximity effects, and magnetically induced substrate losses.

Writing (7.1)(7.2) in matrix notation one obtains

$$\begin{pmatrix} -Y^C S & D \\ D & -Z^M S \\ J & 0 \end{pmatrix} \begin{pmatrix} v \\ i \end{pmatrix} = \begin{pmatrix} 0 \\ v_{s1} \\ v_{s2} \end{pmatrix} \quad (7.3)$$

The last two rows of the above matrix simply enforce boundary conditions at the input and output terminals, which forces these terminals to equal the impressed voltage. Note the right-hand side of the above matrix contains $2N$ zero terms followed by the impressed voltages. The matrix S simply averages whereas the matrix D subtracts adjacent node voltages and terminal currents. In the continuous limit these matrix operators represent integration and differentiation respectively.

Note that the above formulation is fairly general and can include several structures shown in Figures 2.8, 2.9, 2.10, and 2.11. This analysis also naturally applies to series connected multi-layer spirals. The approach can also be applied to the case of multi-layer shunt-connected spirals by simply applying the above technique to "super" segments as opposed to segments, where each "super" segment consists of one or more segments connected in shunt.

4.2.1 CAPACITORS AND DISJOINT INTERCONNECTION OF SEGMENTS

The above procedure can be easily extended to capacitors in the following manner. Introduce a fictitious element so that the plates of the capacitor are connected in series through this element. The electrical properties of this fictitious element should not significantly alter the overall behavior of the device. For instance, a large physical resistor could represent this element. In fact, such a representation is not at all fictitious since the oxide is lossy. But this loss term has already been accounted for by the Green function so we still need to make this resistor value larger than any physical resistor.

This concept can be extended into any geometrically disjoint group of conductors as long as we treat the collection of conductors as a two-port element. The analysis of the previous section yields two-port parameters.

4.2.2 TWO-PORT TRANSFORMER

Another simple modification to the above procedure can be used to analyze a transformer. If the transformer primary and secondary share a common ground, as shown in Fig. 7.4a, then the disjoint primary and secondary are effectively joined at the ground point. If we thus consider the primary and secondary as a

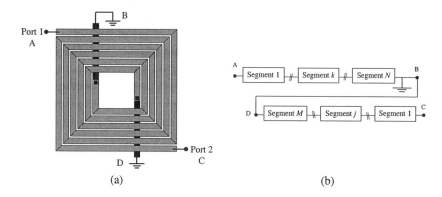

Figure 7.4. (a) A grounded transformer. (b) The primary and secondary connected back-to-back with center point grounded.

single entity, connected back to back, and set up the system equations as shown above, we simply need to shunt one point in the capacitance matrix to ground as shown in Fig. 7.4b to obtain the two-port parameters.

4.3 THREE-PORT TRANSFORMER

Consider two groups of series interconnected segments, as shown in Fig. 2.15. This structure is typically used to form baluns, as shown in Fig. 7.5. The "primary" port consists of l segments, whereas the "secondary" port has $m+n$ segments. The secondary segment is grounded at some arbitrary point, creating two ports consisting of m and n segments, each referenced to ground.

Numbering the nodes as suggested in the figure, we can take the difference between node voltages to obtain

$$\tilde{D}\mathbf{v} = Z^M \mathbf{i} \tag{7.4}$$

The vector $\mathbf{i}$ represents the currents through each segment in the direction shown in the figure. The matrix $\tilde{D}$ is a "punctured" version of the matrix D

$$D_{ij} = \begin{cases} 1 & j = i \\ -1 & j = i+1 \\ 0 & \text{otherwise} \end{cases} \tag{7.5}$$

where the following two locations of D are reset to form $\tilde{D}$

$$\tilde{D}_{l,l+1} = 0 \tag{7.6}$$

$$\tilde{D}_{l+m,l+m+1} = 0 \tag{7.7}$$

to account for the grounded points in the structure. Next, the displacement current is taken into account

$$\tilde{D}^T \mathbf{i} = \tilde{Y}^C \mathbf{v} + \mathbf{i_s} \tag{7.8}$$

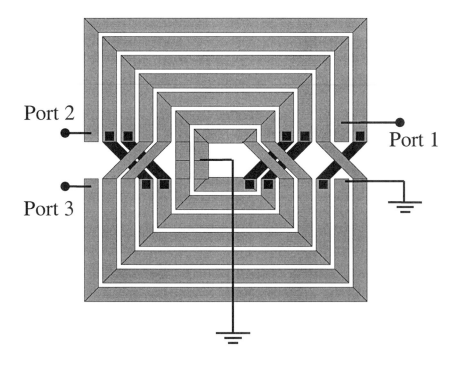

Figure 7.5. A three-port balun.

since a difference between segment currents results in an increase in the charge stored on a segment. The matrix Y^C is the capacitance matrix computed from the partial capacitance matrix as follows

$$\tilde{Y}_{ii} = \sum_{j=1}^{N} Y_{ij} \qquad (7.9)$$

and

$$\tilde{Y}_{ij} = -Y_{ij} \qquad (7.10)$$

The vector $\mathbf{i_s}$ represents the impressed currents. We assume that external current can only enter the device through three terminals shown in the above figure so in general we have the following form

$$\mathbf{i}_{sk} = \begin{cases} -\hat{i}_1 & k = 1 \\ -\hat{i}_2 & k = l+1 \\ -\hat{i}_3 & k = l+m+1 \\ 0 & \text{otherwise} \end{cases} \qquad (7.11)$$

Thus, we have the following system of equations similar to (7.3)

$$\begin{pmatrix} \tilde{D} & -Z^M \\ -\tilde{Y} & \tilde{D}^T \end{pmatrix} \begin{pmatrix} v \\ i \end{pmatrix} = \begin{pmatrix} 0 \\ i_s \end{pmatrix} \quad (7.12)$$

4.4 VISUALIZATION OF CURRENTS AND CHARGES

Note that in the above procedures we went through two discretization steps: one to divide the device into N lumped elements, and then a second step into sub-elements of constant charge and current.

From a two-port network point of view, the detailed current and charge distribution are unimportant. We thus average over the sub-elements when solving the linear equations. It is desirable, though, to view the current and charge distribution at the sub-element level. For instance, this gives us insight into the non-uniform current distribution at different locations in the spiral, such as current constriction at the inner turns. We can do this as follows

$$\tilde{\mathbf{i}} = \tilde{Z}^{M^{-1}} S^T \mathbf{v} \quad (7.13)$$

$$\tilde{\mathbf{q}} = \tilde{Z}^C S^T \mathbf{v} \quad (7.14)$$

where $\mathbf{v}$ is the voltage distribution along the length of the device and the vectors $\tilde{\mathbf{i}}$ and $\tilde{\mathbf{q}}$ give the current and charge distribution along both the long and width/thickness of the device.

Another approach is to solve the system of equations at the sub-element level directly

$$D\tilde{\mathbf{v}} = -\tilde{Z}^M \tilde{\mathbf{i}} \quad (7.15)$$

$$D^T \tilde{\mathbf{i}} = j\omega \underbrace{\tilde{Z}^C \tilde{\mathbf{v}}}_{\text{charge}} + \mathbf{i}_s \quad (7.16)$$

solving for the voltage from the second equation

$$\tilde{\mathbf{v}} = \frac{1}{j\omega} \left(Z^C \right)^{-1} (D^T \tilde{\mathbf{i}} - \mathbf{i}_s) \quad (7.17)$$

Note that the matrix $(Z^C)^{-1} \equiv P$ does not in fact need to be inverted as it is already calculated in this form from the Green function. Substituting in the second equation we have

$$A\tilde{\mathbf{i}} = (DPD^T + j\omega Z^M)\tilde{\mathbf{i}} = DP\mathbf{i}_s \quad (7.18)$$

The matrix A is symmetric and inversion yields the current distribution. The voltage distribution is also given by

$$\tilde{\mathbf{v}} = -D^T Z^M \tilde{\mathbf{i}} \quad (7.19)$$

Note that the D matrix is not the same as before

$$D_{ij} = \begin{cases} 1 & j = i \\ -1 & j = i + S_i \\ 0 & \text{otherwise} \end{cases} \qquad (7.20)$$

where S_i represents the number of sub-elements of segment i.

Chapter 8

EXPERIMENTAL STUDY

In this chapter we will compare measured device characteristics to simulated predictions based on the techniques described in the previous chapters. In a previous report [Niknejad, 1997], experimental results confirmed the validity of the our approach. Several structures were fabricated and measured in a BiCMOS process including square and polygon spiral inductors, planar and non-planar transformers, and coupled inductors.

In this chapter, we would like to make the same comparisons but with devices fabricated over a conductive substrate. As previously discussed, a heavily conductive substrate gives rise to new loss mechanisms, in particular magnetically induced eddy current losses. Conductive substrates are typically employed in CMOS processes to minimize latch-up. Latch-up occurs when the parasitic lateral and vertical bipolar transistors of a CMOS inverter turn on in a positive feedback loop [Johns and Martin, 1997]. To minimize voltage drops in the substrate capable of potentially forward biasing these bipolar devices, a heavily conductive highly doped bulk substrate is employed. A more resistive, lightly doped layer is then grown epitaxially over this substrate to house the wells and MOS transistors, hence the name "epi" substrate. A cross-section of such a process is shown in Fig. 2.5. Another variation, similar to the bipolar substrate, is to start with a lightly doped substrate but to grow a heavily conductive surface layer to shunt substrate currents to a low potential.

1. MEASUREMENT RESULTS

As shown in Fig. 8.1, several planar and non-planar spiral inductors have been fabricated in National Semiconductor's 0.25 μm CMOS-8 process. This process utilizes a bulk substrate of 10 Ω-cm, sufficiently resistive that eddy currents play a minor part in the bulk. However, the top layer of Si is fairly conductive at 15×10^{-4} Ω-cm. The thickness of this layer is less than 1 μm but

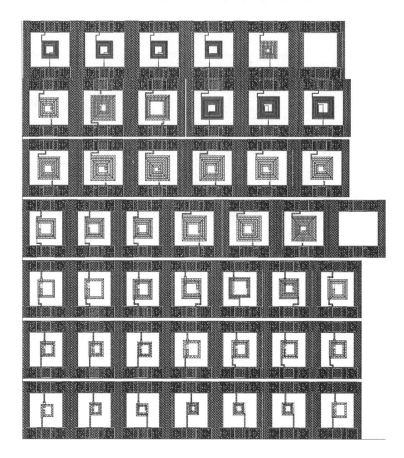

Figure 8.1. Inductor test structures.

this is enough to cause significant eddy current loss. The process parameters are summarized in Table 8.1.

The layout of two representative inductors is summarized in Table 8.2. As shown in Fig. 8.2, spiral inductor L27 is a planar device utilizing the top metal layer. Inductor L19 resides on metal layers M5-M3 connected in series realizing a large inductance value in a relatively small area. The layout of L19 is shown in Fig. 8.3.

2. DEVICE CALIBRATION

Measurements are performed using a ground-signal-ground co-planar waveguide pad configuration. The s-parameters are measured using the HP 8719C Network Analyzer. G-S-G co-planar cascade probes are used and the setup is calibrated using the Cascade Microtech 832210 calibration substrate. The

Table 8.1. CMOS Process Parameters Summary

Metal 5	$R_{sh} = 35\, m\Omega/sq$	$t = 0.91\mu m$	$C_{sub} = 6.12 aF/\mu m^2$
			$C_{M4} = 53.1 aF/\mu m^2$
			$C_{M3} = 19.1 aF/\mu m^2$
Metal 4	$R_{sh} = 60\, m\Omega/sq$	$t = 0.51\mu m$	$C_{sub} = 7.69 aF/\mu m^2$
			$C_{M3} = 53.1 aF/\mu m^2$
Metal 3	$R_{sh} = 60\, m\Omega/sq$	$t = 0.51\mu m$	$C_{sub} = 10.4 aF/\mu m^2$
Surface Layer	$\rho = 15 \times 10^{-4}\, \Omega-cm$	$t = 0.8\mu m$	p^+ Si
Bulk Substrate	$\rho = 10\, \Omega-cm$	$t = 725\mu m$	p^- Si

Table 8.2. Device Physical Dimensions

Spiral Name	L19	L27
Outer Length (μm)	200	250
Metal Width W (μm)	11	10.5
Metal Spacing S (μm)	3	3
No. of Turns N	2.5	7.75
Metal Layer(s)	M3-M5	M5

open-pad y-parameters are also measured and subtracted from the measured y-parameters to remove the pad capacitance and loss.

It is interesting to note that the G-S-G approach has some clear advantages over the S-G approach presented previously in [Niknejad and Meyer, 1998]. To see this, consider the ground pad configurations of each approach, as show in in Fig. 8.4. As noted in [Niknejad, 1997], the S-G approach is prone to calibration errors resulting from the substrate coupling between the pads, since the presence of the device under test alters the substrate coupling. On the other hand, in the G-S-G approach, the substrate pads are widely separated and the presence of the ground de-couples the input from the output port. Thus, the G-S-G calibration step only subtracts out the input and output capacitance and has a minor influence on the coupling capacitance between the input and output ports, in stark contrast to the S-G approach where a large coupling capacitance is involved.

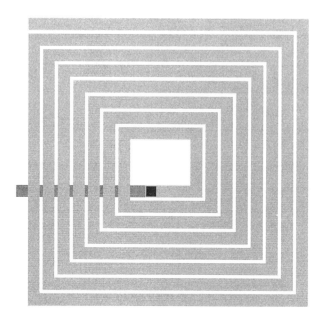

Figure 8.2. Layout of inductor L27.

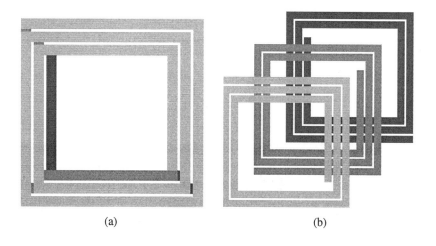

Figure 8.3. Layout of inductor L19. In part (b) the lower metal layers have been staggered to give a clear picture of the device geometry.

For these reasons, the G-S-G calibration procedure was selected. Designing a proper G-S-G calibration test structure in CMOS requires special care. With a device under test connected to the calibration structure, return ground currents flow through the calibration structure. To minimize the losses resulting

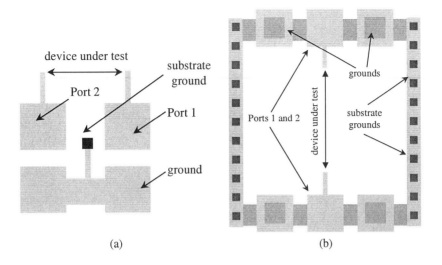

Figure 8.4. Calibration pad structures. (a) S-G structure. (b) G-S-G structure.

from such ground currents, metal layers M1-M5 have been strapped with an ample supply of vias and substrate ground connections to minimize the losses. Furthermore, in the G-S-G approach the signal ports can be isolated from the substrate by using an M1 shield under the pads, minimizing the loss contribution from the pads. In the S-G case, the pads introduce substantial loss which must be calibrated out of the device loss.

3. SINGLE LAYER INDUCTOR

Measured s-parameters for inductor L27 are shown in Fig. 8.5. The simulated and measured results match well. The discrepancy above the self-resonant frequency is in the capacitive region where we are less interested in the device. Notice that the inductor self-resonates at a frequency of 4.25 GHz. Simulations using *ASITIC* predicted a self-resonant frequency of 4.15 GHz. The simulations are performed on a Pentium II 400 MHz machine running the Linux operating system. Each frequency point requires less than ten seconds of computation. In Fig. 8.6 we plot the effective value of inductance. This is derived using a one-to-one transformation of the s-parameters into π-parameters [3]. Again, a good match is observed between the theory and measurements. The inductance decrease is due mostly to the capacitive effects rather than the inductive effects. Inductance value decreases slightly due to skin effect and eddy currents in the substrate but the main reason for the decrease is that energy is coupled from port to port through the winding capacitance at higher frequencies.

Fig. 8.7 shows the effective value of series resistance as a function of frequency. Two simulations are performed, with and without eddy current losses.

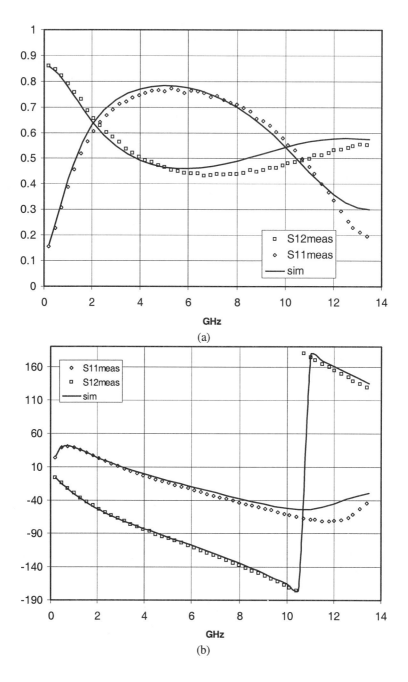

Figure 8.5. Measured and simulated (a) magnitude and (b) phase *s*-parameters of spiral inductor L27.

Experimental Study 115

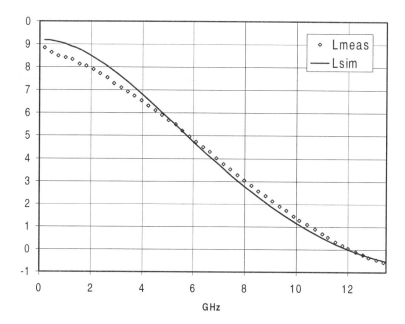

Figure 8.6. Measured and simulated inductance (imaginary component of Y_{21}) of spiral inductor L27.

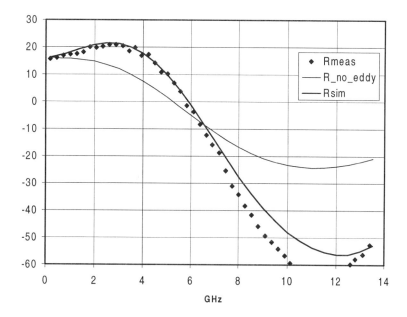

Figure 8.7. Measured and simulated resistance (real component of Y_{21}) of spiral inductor L27.

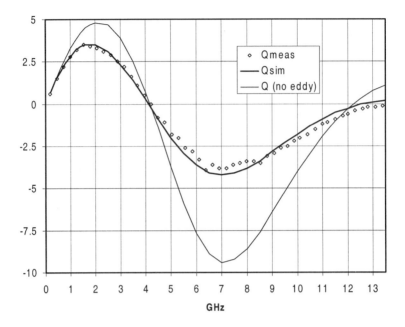

Figure 8.8. Measured and simulated quality factor (imaginary over real component of Y_{21}) of spiral inductor L27.

As evident in the figure, eddy current losses are critical to model. The variation in frequency of the series resistance is due to various competing effects. Skin effect and proximity effects increase the series resistance but beyond 1 GHz this is swamped by the increase from eddy currents. At higher frequencies more energy is transported capacitively, and consequently the resistance decreases and eventually becomes negative. The real part of the total input impedance looking into each port, of course, is positive at all frequencies.

The quality factor Q, the ratio between the imaginary and real part of the input impedance, is plotted in Fig. 8.8. Again, a good match is observed between the theory and measurements. Note that negative quality factor implies that the device is acting as a capacitor rather than an inductor. In reality, this plot is misleading as it implies a Q of zero at self-resonance. A better way to calculate Q is given in [3] but for comparison the given definition is better since it involves a minimal transformation of the measured s-parameters. The substrate resistance and capacitance are also shown in Fig. 8.9 and Fig. 8.10. The overall shape of both curves matches the measurements well. The low frequency substrate resistance measurements are noisy due to measurement error.

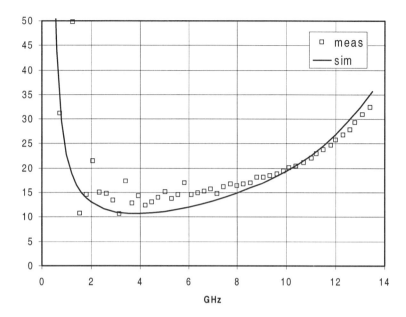

Figure 8.9. Measured and simulated substrate resistance of spiral inductor L27.

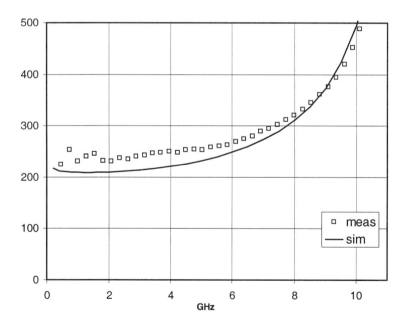

Figure 8.10. Measured and simulated substrate capacitance of spiral inductor L27.

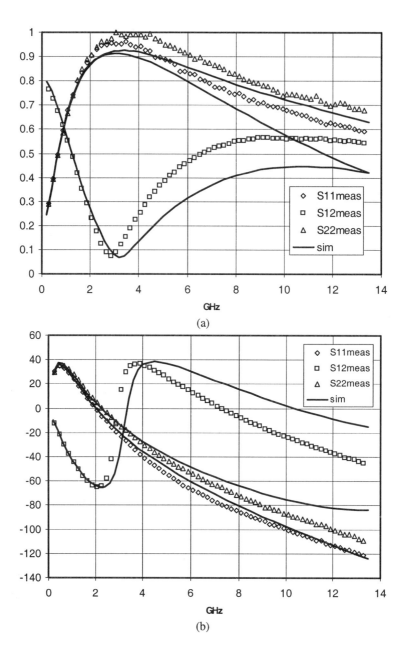

Figure 8.11. Measured and simulated (a) magnitude and (b) phase s-parameters of spiral inductor L19.

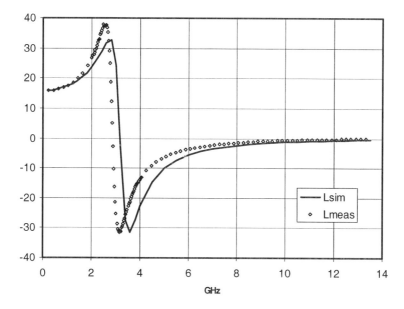

Figure 8.12. Measured and simulated inductance (imaginary component of Y_{21}) of spiral inductor L19.

4. MULTI-LAYER INDUCTOR

Measured s-parameters for inductor L19 are shown in Fig. 8.11. A fairly good match between the simulated and measured is observed, especially below the self-resonant frequency. Due to the large interwinding capacitance intrinsic to this series-connected structure, the self-resonant frequency is low. The self-resonant frequency is predicted at 2.57 GHz and measured at 2.47 GHz.

To gain insight into the data we also plot the series inductance and resistance in Figures 8.12 and 8.13. These curves are dramatically different from the previous device. The differences can be accounted for by noting that a planar device self-resonates through the substrate capacitance whereas a multi-metal device self-resonates through the interwinding capacitance. Since this is a relatively high Q capacitor compared to the lossy substrate capacitance, the behavior of L and R can be modeled by the simple equivalent circuit of Fig. 8.14. As we approach self-resonance the inductance and resistance values peak. The maximum value of resistance is approximately $Q^2 R_{\text{series}}$. Thus, the lower the losses the higher the value of the peak. Again, simulation is able to predict these effects well. The discrepancy is due mainly to the value of oxide thickness used in the simulation. The process parameters were not adjusted at all in the simulations in order to gauge how well one can predict inductor performance *a priori*. Since oxide thickness is not a well-controlled process parameter, and since the oxide thickness has a significant impact on the device performance, it

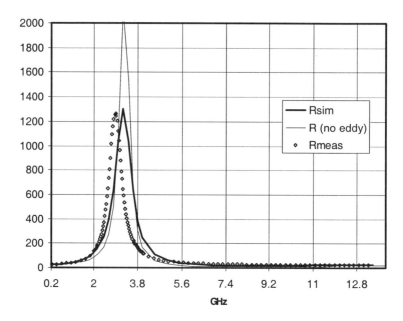

Figure 8.13. Measured and simulated resistance (real component of Y_{21}) of spiral inductor L19.

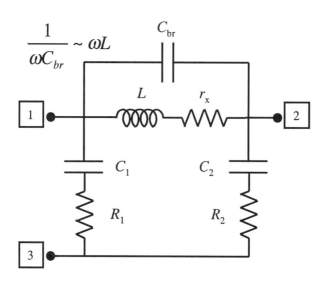

Figure 8.14. Simple equivalent circuit of inductor L19 close to self-resonance.

seems like this device is not as manufacturable as the planar structure, especially close to self-resonance.

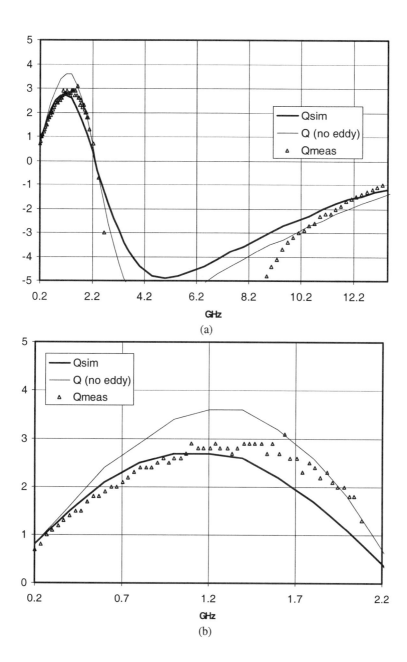

Figure 8.15. (a) Measured and simulated quality factor (imaginary over real component of Y_{21}) of spiral inductor L19. (b) Close-up plot.

Fig. 8.15 shows the quality factor plot for two simulations and the measurements. As before, the simulation results with no eddy currents under-predict the losses, but now the simulations with eddy currents slightly over-predict the losses. This is especially true beyond self-resonance where the device is acting as a capacitor. Fortunately, one is generally not interested as much in frequencies beyond self-resonance. The discrepancy, though, can be easily explained with some physical insight into the problem. Beyond self-resonance the device is acting very much like a metal-insulator-metal capacitor. Since wide metal lines are used to wind the inductor to minimize the low frequency losses, the current distribution in the lines is not constrained to be one-dimensional as we have assumed. In fact, at high frequencies the bottom metal layers act as a solid shield against both the magnetic and electric fields. Thus, eddy currents flow in the shield and prevent the magnetic fields from penetrating the lossy substrate. We could account for this effect by utilizing a two-dimensional mesh but we are less interested in this region of operation.

II
APPLICATIONS OF PASSIVE DEVICES

Chapter 9

VOLTAGE CONTROLLED OSCILLATORS

1. INTRODUCTION

Consider the simplified block diagram of a superheterodyne transceiver, shown in Fig. 9.1. The voltage controlled oscillator (VCO) is a key components of this transceiver as the VCO generates the 'local oscillator' or LO frequency. The LO in turn drives the receive and transmit mixers, converting the received signal from RF to IF or baseband and similarly converting the baseband and IF

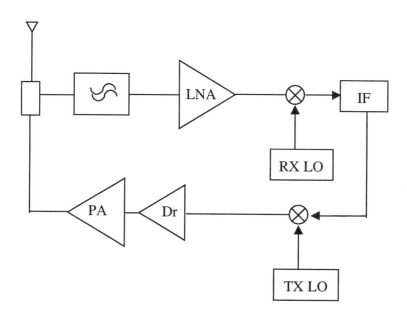

Figure 9.1. Simplified block diagram of a superheterodyne transceiver front-end.

signals to RF for transmission. This conversion process, or mixing, is achieved through multiplication of the sinusoidal output of the VCO with the modulated signal. The LO signal is given by

$$v_{LO}(t) = A\cos(\omega_0 t + \phi_n(t)) \tag{9.1}$$

where A is the constant amplitude of the oscillator, $\omega_0 = f(V_{\text{control}})$ is the frequency of oscillation which can vary by application of the control voltage, and $\phi_n(t)$ represents the random phase noise of the oscillator.

Now consider the information-containing stochastic bandpass signal, $x(t)$, with spectral components centered around ω_1. Such a process can be uniquely decomposed into two parts as follows [Proakis, 1995]

$$x(t) = B(t)\cos(\omega_1 t + \phi(t)) \tag{9.2}$$

$B(t)$ represents the amplitude modulation and $\phi(t)$ the phase modulation of the signal. Multiplication of the modulated signal with the LO

$$\begin{aligned} v_o(t) &= [A\cos(\omega_0 t + \phi_n(t))] \times [B(t)\cos(\omega_1 t + \phi(t))] \\ &= \frac{AB(t)}{2}[\cos((\omega_0 - \omega_1)t + \phi_n(t) - \phi(t)) + \\ &\quad \cos((\omega_0 + \omega_1)t + \phi_n(t) + \phi(t))] \end{aligned} \tag{9.3}$$

Note that the conversion process has moved the spectral energy from around ω_0 to $\omega_0 \pm \omega_1$. In the case that $x(t)$ represents the received RF signal, then the conversion process moves the input spectrum to a more convenient frequency, usually a lower frequency. In the case that $x(t)$ represents the baseband signal, the conversion process moves the input spectrum to a frequency more convenient for transmission, usually a higher frequency.

Since antenna efficiency is directly related to antenna size relative to the wavelength [Jackson, 1999] at the frequency of interest, higher frequencies allow smaller and more portable antennas to be realized without sacrificing efficiency. In the case of a wired link, mixing with the LO is still useful if the transmission media is shared by several transceivers. In such a case frequency division multiplexing (FDM) or frequency hopping multiple access (FHMA) can be used to share the channel. As evident from (9.1), the phase information can be modulated onto the LO signal directly by means of the voltage control line.

In the superheterodyne receiver, the conversion process occurs in two stages, and the first conversion moves the received signal to the intermediate frequency, or IF. The IF frequency is chosen to simplify the filtering and detection of the input signal. Typically, the input signal bandwidth is much smaller than the carrier signal and thus filtering at RF requires very high Q components. Hence, there is great incentive to perform filtering and channel selection at IF rather than at RF directly. To alleviate the need for a variable frequency bandpass filter, the LO signal is adjusted so that the IF frequency $\omega_0 - \omega_1$ is a constant center frequency allowing a fixed frequency filter to perform channel selection at IF.

The detrimental effects of the down-conversion or mixing operation are evident when we consider phase noise of the LO

$$V_{IF}(t) = cos((\omega_0 - \omega_1)t + \phi_n(t) - \phi(t)) \qquad (9.4)$$

Note that for proper detection of phase modulation we require that $|\phi_n(t)| \ll |\phi(t)|$. Even when this condition is met, though, the occurrence of a third interfering signal at the input of the receiver can introduce "reciprocal mixing" which can greatly degrade the in-band SNR [Razavi, 1998].

On the transmit side, a low-frequency information signal is modulated onto a higher frequency carrier appropriate for radiative (or optical) transmission. In such a case, the phase noise of the LO leaks outside of the channel bandwidth and through amplification and radiation in the transmit chain appears at the input of transceivers receiving in neighboring bands. This can potentially block or degrade the SNR performance of other users of the spectrum.

2. MOTIVATION

VCOs, as key building blocks in wireless transceivers, require careful design and optimization. The principal performance specifications are power consumption and phase noise performance. Monolithic VCO implementations suffer from poor phase noise performance partly due to low quality factor Q passive components in the frequency band of interest 1–2 GHz. The inductors at these frequencies usually dominate the tank Q. At higher frequencies, 3–10 GHz, it is possible to obtain higher Q inductors over a moderately resistive substrate.

Thus there is a motivation to implement VCOs at higher frequencies where higher Q inductors are easier to fabricate. On the other hand, the varactor Q decreases with frequency and most processes have varactors optimized for 1–2 GHz. Hence, there exists an optimum frequency where the total tank Q is maximum.

While VCOs at higher frequencies can be used directly in many upcoming wireless standards such as wireless LANs at $\sim$ 5 GHz, these VCOs can also be applied to traditional standards in the 1–2 GHz range by the use of frequency dividers. Dividing the VCO output improves the phase noise with little added power consumption. Using a different LO frequency also mitigates coupling problems inherent in monolithic transceiver designs, such as the power amplifier locking the VCO.

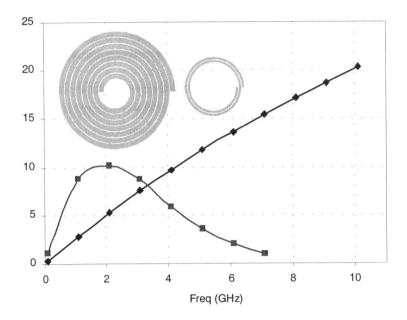

Figure 9.2. The simulated quality factor of a small footprint 1 nH inductor versus an optimal 10 nH inductor. The smaller inductor dimensions are $R = 75\mu m$ $W = 5\mu m$ $N = 2$ while the larger inductor has dimensions $R = 150\mu m$ $W = 12.3\mu m$ $N = 7.5$. Both structures are realized with a metal spacing $S = 2.1\mu m$.

3. PASSIVE DEVICE DESIGN AND OPTIMIZATION
3.1 INDUCTOR LOSS MECHANISMS

The analysis and characterization of inductors and other passive devices is a key prerequisite for successful VCO design. Since the VCO phase noise performance is highly dependent on the Q of the tank, as shown approximately from the well-known Leeson formula [Leeson, 1966], it is desirable to obtain high Q inductors.

As we have seen, inductor Q is limited by physical phenomena that convert electromagnetic energy into heat or radiation. If the substrate is sufficiently conductive, magnetically induced currents, or bulk eddy currents, flow in the substrate and act as a possibly dominant form of loss. In the case of highly conductive substrates, such as those of an epi CMOS process, eddy current losses indeed severely limit the Q. In the case of a moderately conductive substrate, such as a bipolar or BiCMOS substrate with resistivity $\sim 10\ \Omega$-cm, the bulk substrate loss mechanisms are dominated by electrically induced substrate currents while magnetically induced substrate currents are negligible even up to 10 GHz.

Therefore for such moderately conductive substrates there is a great benefit in designing higher frequency VCOs. Substrate losses are curtailed by scaling the inductor size. This also saves valuable chip area since capacitors scale in size at higher frequencies. Skin effect and proximity effect losses prevail but careful analysis can minimize these effects. As a comparison, consider the Q of a 1 nH inductor at 10 GHz versus a 10 nH inductor at 2 GHz. As shown in Fig. 9.2, the simulated maximum quality factor of the 1 nH inductor far exceeds the 10 nH inductor with a considerable savings in chip area.

3.2 DIFFERENTIAL QUALITY FACTOR

It has been recognized that the quality factor of an inductor at high frequency is higher seen differentially rather than single-endedly [Kuhn et al., 1995] [Danesh et al., 1998]. This can be deduced in the following way: if we consider an inductor as a series interconnection of a set of coupled transmission lines [Shepherd, 1986], then we can model the inductor by an equivalent transmission line of appropriate length and impedance by diagonalizing the transmission line matrix. Intuitively, one can see that a shorted inductor self-resonates fundamentally at the quarter-wavelength frequency whereas a differential transmission line self-resonates at the half-wavelength frequency. Hence, we expect an approximate factor of two improvement in the self-resonant frequency of a differential inductor. This means that at a given frequency, less current is injected capacitively into the substrate and therefore fewer substrate losses occur.

It follows that VCO topologies that inherently employ the inductor differentially are advantageous. Differential circuits naturally fit this definition. Using symmetric center-tapped inductors as opposed to two uncoupled inductors leads to a savings in chip area due to the mutual magnetic coupling. Furthermore, the savings in area leads to higher Q values since at high frequencies substrate losses dominate the Q. Lastly, there is no need to model the parasitic coupling that occurs between the two inductors [Meyer et al., 1997].

In this design we employ the center-tapped layout shown in Fig. 9.3 as opposed to the design presented in [Kuhn et al., 1995]. This layout has also been used by [Danesh et al., 1998] [Long and Copeland, 1997]. Fig. 9.4 shows that the maximum Q enhancement is almost a factor of two differentially over single-ended drive. A broadband 3-port equivalent circuit model, shown in Fig. 9.5, can be derived using *ASITIC*.

3.3 VARACTOR LOSSES

A high quality varactor is needed to achieve low phase noise over the full tuning range of the VCO. The varactor is usually designed as a reverse biased PN junction diode. The varactor Q is limited by the intrinsic series resistance

130   INDUCTORS AND TRANSFORMERS FOR SI RF ICS

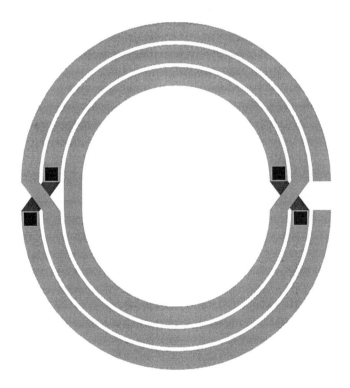

Figure 9.3. Layout of a center-tapped spiral inductor with radius $R = 125$ μm and the metal width $W = 14.5$ μm.

of the device. Given a layout geometry, *ASITIC* can be used to calculate this distributed resistance.

Typically the varactor Q is sufficiently high in the 1–2 GHz region and thus the varactor loading on the tank is negligible. At increasingly higher frequencies, though, it becomes more difficult to realize a high Q varactor since the $Q = (\omega R_x C_v)^{-1}$ decreases as a function of frequency. To circumvent this, some researchers advocate electrostatically tuned varactors [Young and Boser, 1197] [Dec and Suyama, 1997]. In our process, optimized varactors for 1 GHz were available. These varactors utilize *npn* transistors with higher levels of doping to minimize intrinsic base and emitter resistance. To minimize the noise injection from the control line, a differential structure was used, as shown in Fig. 9.6. The emitters are tied to the tank to allow a monolithic realization since for a single supply design the control voltage range must not exceed the supply voltage. To avoid forward biasing the junctions, the control line voltage V_C should not exceed $V_C \leq V_{CC} - v_o$ where v_o is the steady state oscillation amplitude.

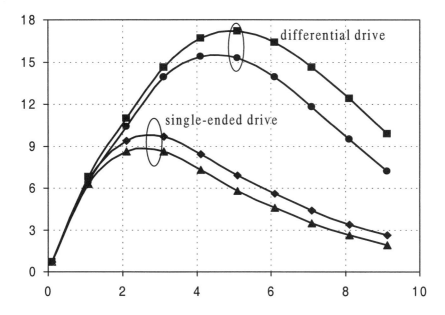

Figure 9.4. Simulated quality factor driven single-endedly (bottom curve) versus differentially (top curve). Note that the comparison is made with both a circular (higher Q value) and square device.

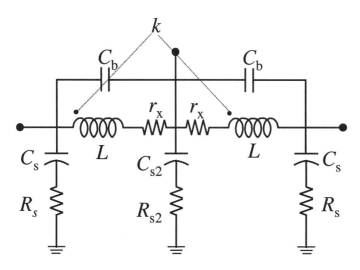

Figure 9.5. Compact circuit model for center-tapped spiral inductor.

To first-order, since the common base connection is a virtual ground for differential signals, noise injected at the control line is rejected. The time-varying nature of the reverse bias voltage, though, leads to noise appearing at

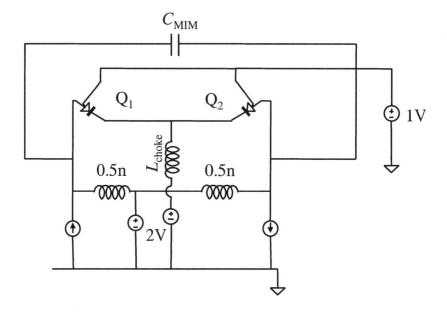

Figure 9.6. Schematic of differential varactors.

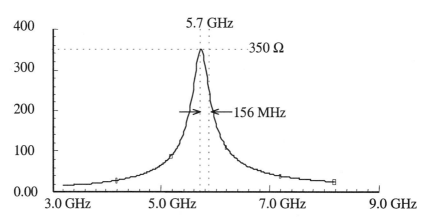

Figure 9.7. Simulated varactor resonance curve.

the output. To avoid this, care must be exercised to properly bypass the control point to ground. Also, the collector node of the varactor is connected to V_{CC} and not to the base to isolate the collector losses from the tank.

The resonance curve of the differential varactor is shown in Fig. 9.7. Note that the varactor Q is on the same order of magnitude as the inductor Q. Hence, to avoid loading the tank it is necessary to add linear capacitance to the tank to increase the effective Q.

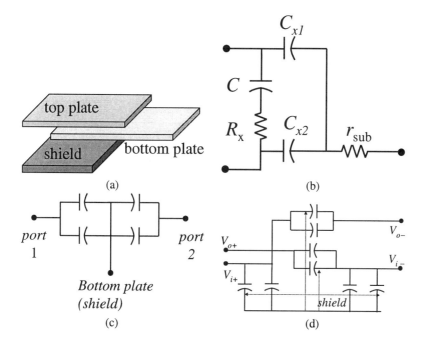

Figure 9.8. Shielded MIM capacitor equivalent circuit. In (a) the top, bottom, and shielding plates are shown. Part (b) shows the equivalent circuit extracted from *ASITIC*. In (c) the load capacitor is shown realized as a symmetric-quad. In (d) the passive capacitive feedback network is shown, also realized as a pair of shielded symmetric quads.

3.4 METAL-INSULATOR-METAL (MIM) CAPACITORS

To realize a high quality factor for the entire tank, high Q capacitors are needed to boost the effective quality factor of the varactor. One may employ a linear high Q capacitor in series or in shunt with the varactor. Both connections limit the tuning range of the VCO. The series connection has the added advantage of allowing a greater range of tuning voltage to be applied to the tank without forward biasing the varactor diode.

While MOS capacitors were an option in this process, we opted for MIM capacitors to avoid any further loading of the tank. MIM capacitors have low density and thus require larger chip area. The process variation of the oxide thickness also further limits the applicability of MIM capacitors in Si. Since MIM capacitors only make up a fraction of the total tank capacitance, the variation in oxide thickness is tolerable.

To avoid substrate injection and the consequent substrate losses, the MIM capacitors are shielded with the bottom metal layer. An equivalent circuit, calculated using *ASITIC*, is shown in Fig. 9.8. To avoid the losses of the bottom plate parasitics of the shielded capacitors, a differential structure is employed with the bottom plate at the virtual ground.

4. VCO CIRCUIT DESIGN
4.1 VCO TOPOLOGY

For this work the differential topology is a natural choice given the discussion in Section 3.2. Differential operation has many other positive attributes amenable to monolithic integration. It provides better immunity from the package and substrate parasitics, reducing substrate coupling. This is especially important when the VCO is integrated with many other blocks. Differential dividers are also easier to realize at high frequencies as emitter coupled logic can be employed. While the area of active devices is doubled, the area of passive devices reduces due to the mutual coupling and higher differential Q.

For a generic differential pair with positive feedback, the equivalent negative resistance is given by

$$R_{eq} = \frac{-2n}{g_m} \tag{9.5}$$

where the factor n is the fraction of voltage fed back to the base of the differential pair. As shown in Fig. 9.9 the feedback network can be realized using an on-chip transformer [Zannoth et al., 1998], emitter followers [Jansen et al., 1997], a direct connection [Razavi, 1998], or a capacitive transformer [Razavi, 1998]. We opted for a capacitive transformer in order to maximize the oscillation amplitude while minimizing the loading on the tank, the power dissipation, and the noise.

4.2 PHASE NOISE ANALYSIS

Full characterization of the phase noise requires numerical solutions to stochastic differential equations [Kaertner, 1990, Demir et al., 1998]. Approximate linear time-varying techniques have been presented by [Hajimiri and Lee, 1998, Samori et al., 1998] and agree well with measurements. These techniques are more appropriate for simulation rather than the design of a VCO.[1] To gain an understanding of the operation of the VCO and phase noise, we follow the technique presented in [Kouznets and Meyer,]. This is a simple engineering approach that gives good insight into the problem.

4.2.1 LINEAR ANALYSIS NOISE TRANSFER COEFFICIENTS

Consider Fig. 9.11, the simplified small signal equivalent circuit of Fig. 9.10. For simplicity, C_μ and r_π have been neglected in the equivalent circuit. Current sources $i_{1a,b}$ model the shot noise generated by the differential pair collector junction. Current sources $i_{2a,b}$ model the base current shot noise. Note that while we ignore the intrinsic base resistance r_b, it is critical to include the noise

Voltage Controlled Oscillators 135

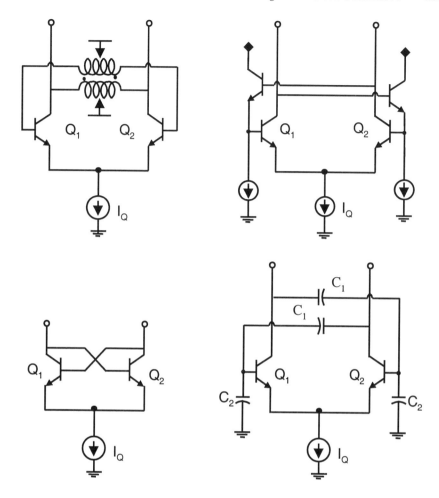

Figure 9.9. Different feedback mechanisms for negative resistance generation.

generated by r_b, and this is represented by the voltage sources $v_{3a,b}$. The current sources $i_{4a,b}$ represent the current noise generated by the biasing resistors $R_{B1,2}$. Current source i_5 represents the total noise generated by the passive tank, including the thermal noise of the metals that make up the inductance as well as the substrate losses.

From a linear time-invariant point of view, our noise description of the circuit is complete. Due to symmetry, the common emitter node is a virtual ground and thus the noise due to the tail current is shunted to ground. Furthermore, due to the differential output, any noise injected at the emitters is a common mode signal that is attenuated by the common mode rejection, set by the level of matching in the circuit. This argument, though, neglects the time variance of

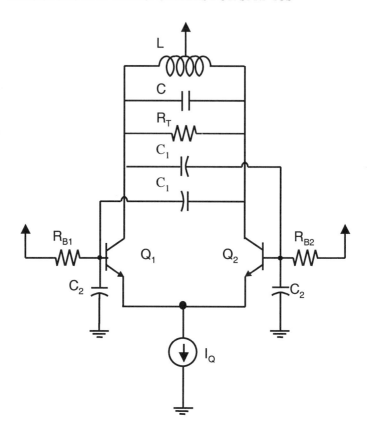

Figure 9.10. Schematic of differential VCO.

the circuit which results in a time-varying transfer function from the tail current source to the output.[2]

Writing the KVL equations at the input base nodes we have

$$v_{i1,2}sC_{2,4} - i_{4a,b} - i_{2a,b} + (v_{i1,2} - v_{3a,b})sC_{\pi 1,2} + (v_{i1,2} - v_{o2,1})sC_1 = 0 \quad (9.6)$$

similarly at the output collector nodes

$$\begin{aligned} 0 = {} & g_{m1,2}(v_{i1,2} - v_{3a,b}) - i_{1a,b} + +(v_{o1,2} - v_{o2,1})sC_T + \\ & (v_{o1,2} - v_{o2,1})\frac{1}{sL} + (v_{o1,2} - v_{o2,1})\frac{1}{R_T} \mp i_5 + \\ & (v_{o1,2} - v_{i2,1})sC_{3,1} \end{aligned} \quad (9.7)$$

Solving the above system of equations for the differential output voltage we have

$$v_{od} = (a_{i1}i_{1a,b} + a_{i2}i_{2a,b} + a_{v3}v_{3a,b} + a_{i4}i_{4a,b} + a_{i5}i_5) \quad (9.8)$$

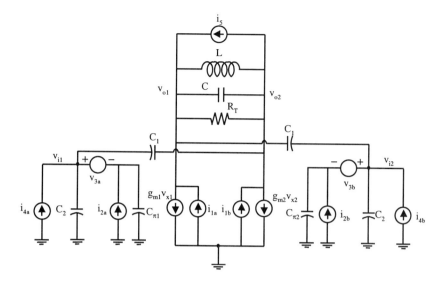

Figure 9.11. Simplified equivalent circuit of differential VCO.

where

$$a_{i1} = \frac{\frac{1}{2}sL}{D(s)} \tag{9.9}$$

$$a_{i2} = \frac{-\frac{1}{2}(g_m + sC_1)\frac{L}{C_\sigma}}{D(s)} \approx \frac{-\frac{1}{2}g_m \frac{L}{C_\sigma}}{D(s)} \tag{9.10}$$

$$a_{v3} = \frac{\frac{1}{2}\frac{C_\pi}{C_\sigma}sL(g_m + sC_1)}{D(s)} \approx \frac{\frac{1}{2}\frac{C_\pi}{C_\sigma}sLg_m}{D(s)} \tag{9.11}$$

$$a_{i4} = a_{i2} \tag{9.12}$$

$$a_{i5} = 2a_{i1} \tag{9.13}$$

where $D(s)$ is the determinant of the system (9.6) (9.6)

$$D(s) = 1 + (\frac{C_1 g_m}{2C_\sigma} - \frac{1}{R_T})sL + LC_{eff}s^2 \tag{9.14}$$

and C_σ represents the total capacitance at the base of each transistor

$$C_\sigma = C_1 + C_2 + C_\pi \tag{9.15}$$

The capacitance C_{eff} is given by

$$C_{eff} = C_T + \frac{C_1(C_2 + C_\pi)}{2C_\sigma} \tag{9.16}$$

In steady state oscillation, the second expression of (9.14) vanishes and the frequency of oscillation is set by

$$\omega_0 = \frac{1}{\sqrt{LC_{eff}}} \qquad (9.17)$$

The vanishing of the second term corresponds to the balancing of the circuit losses with the negative resistance of the differential pair. Equivalently, the balancing occurs when the loop gain of the system is equal to unity. Thus the above equation can be rewritten as

$$D(s) = 1 + (A_\ell - 1)\frac{sL}{R_T} + \frac{s^2}{\omega_0^2} \qquad (9.18)$$

The expression A_ℓ is identified as the loop gain of the system

$$A_\ell = \frac{g_m R_T}{2n} \qquad (9.19)$$

where n is the feedback factor of the capacitive transformer coupling signals from the collector to the base

$$\frac{1}{n} = \frac{C_1}{C_\sigma} \qquad (9.20)$$

In steady state $A_\ell \to 1$ and thus at frequencies close to the oscillation frequency ω_0, (9.18) can be simplified to

$$D(\omega_0 + \delta\omega) \approx \frac{-2\delta\omega}{\omega_0} \qquad (9.21)$$

This results in the characteristic 6 dB/octave decrease in noise power when one moves away from the carrier, far from the flicker noise corner of the device.

4.2.2 NOISE POWER AT OUTPUT

The total power spectral density of the noise at the output of the differential pair can now be written [Gray and Meyer, 1993]

$$\overline{v_{od}^2} = \frac{1}{2}(a_{i5}^2 \overline{i_5^2} + 2a_{i4}^2 \overline{i_4^2} + 2a_{v3}^2 \overline{v_3^2} + 2a_{i2}^2 \overline{i_2^2} + 2a_{i1}^2 \overline{i_1^2}) \qquad (9.22)$$

The factors of two inside the parentheses account for the differential nature of the circuit. The factor of $\frac{1}{2}$ outside of the parentheses is due to the transfer of noise at frequency $(-\omega_0 - \delta\omega)$ to the output [Kouznets and Meyer,].

The noise power spectral densities of the sources are given by [Gray and Meyer, 1993]

$$\overline{i_1^2} = 2qI_C \qquad (9.23)$$

$$\overline{i_2^2} = 2qI_B + K_1 \frac{I_B^a}{f} \quad (9.24)$$

$$\overline{v_3^2} = 4k_B T r_b \quad (9.25)$$

$$\overline{i_4^2} = \frac{4k_B T}{R_B} \quad (9.26)$$

$$\overline{i_5^2} = \frac{4k_B T}{R_T} \quad (9.27)$$

For a typical bipolar technology, the flicker noise corner frequency is on the order of 1–10 kHz and thus the direct contribution of the flicker noise term in (9.24) can be neglected at RF frequencies. On the other hand, time variance in the circuit can translate the noise around DC to RF.

To see how the Q of the tank influences the noise gain, consider the current and voltage gains from the base of the transistors to the output. Assume for simplicity that the inductor Q dominates and for g_m substitute the steady state value that solves $A_\ell = 1$. This leads to

$$|a_{i2}|^2 \approx \frac{1}{4C_1^2 Q^2 \delta\omega^2} \quad (9.28)$$

and

$$|a_{v3}|^2 \approx \frac{C_{pi}}{C_1}^2 \frac{\omega_0^2}{4Q^2 \delta\omega^2} \quad (9.29)$$

This quadratic dependence on Q is well-documented [Razavi, 1998]. The other noise transfer coefficients do not depend directly on Q

$$|a_{i1}|^2 = \frac{(\omega_0 L)^2 \omega_0^2}{\delta\omega^2} \quad (9.30)$$

$$|a_{i5}|^2 = 4|a_{i1}|^2 \quad (9.31)$$

To minimize the above coefficients, one must minimize the tank inductance. Naturally this requires a compensating increase in the tank capacitance. A similar conclusion is reached by [Samori et al., 1998] and can be verified using SpectreRF simulations. In a practical design, the lower limit is set by the tolerance of L and the realizability of C.

4.3 COMPARISON WITH SPECTRERF SIMULATION

Figure 9.12 shows the simulated and calculated phase noise of a simple VCO with parameters summarized in Table 9.1. The simulated phase noise is computed using SpectreRF. The calculated phase noise is from the equations of the previous section. A good match is observed between theory and calculation. On the other hand, close to the carrier, the mixing effects of the fundamental

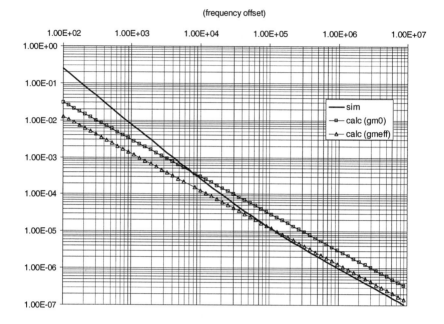

Figure 9.12. Simulated and calculated phase noise spectrum.

Table 9.1. VCO circuit and process parameters.

Bias Current = 5 mA	
$L = 3$ nH	
$C1 = C2 = 200$ fF	$C_T = 856$ fF
$R_T = 600\,\Omega$	$R_B = 30$ kΩ
$r_b = 32.4\,\Omega$	$f_T = 25$ GHz
$\beta_0 = 168$	$C_{je0} = 48$ fF

component are significant, as can be seen from the close-in phase noise below 100 kHz. Here the rise in noise is faster than ω^{-2} due to the flicker noise up-conversion.

A more thorough comparison can be performed if we sweep the oscillator design parameters A_ℓ, Q, and n. Linear analysis predicts the phase noise well when the loop gain is not too large. This conclusion has also been reached by [Kouznets and Meyer,] for the single-ended Colpitts oscillator.

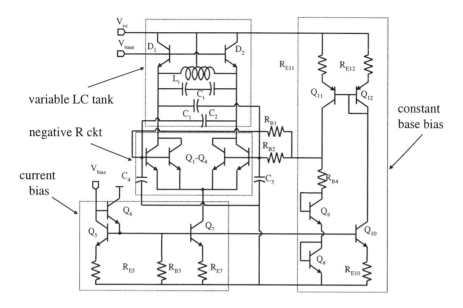

Figure 9.13. VCO core circuit schematic.

5. VCO IMPLEMENTATION

Fig. 9.13 shows a schematic of the VCO core in our practical implementation. Q1-Q4 form a differential quad and are sized each at $5 \times 0.4 \times 0.7 \mu m$. Shielded (MIM) capacitors C1-C4 of value 100 fF (60 fF)^3 are used in the feedback network to allow maximum swing The capacitors are also in quad formation for matching purposes. Frequency tuning is performed via varactors D1-D2. These are reverse-biased base-emitter junctions with extra doping to minimize series resistance. Nevertheless, the quality factor of the varactors is too low at high frequencies, and thus some 300 fF (0 fF) of MIM capacitance is added to the tank. Inductor L_t is a 3 nH (1.8 nH) inductor realized as a center-tapped device as shown in Fig. 9.3. It has a simulated peak Q of 15 at 3 GHz.

To find the optimum bias current provisions are made to vary the bias current externally. If the base current biasing resistors were tied directly to the supply as in Fig. 9.10, the base DC voltage would vary as a function of β as well as a function of collector current and temperature T. This variation could potentially saturate the tail current source, and to avoid this Q5-Q9 are used to provide a relatively constant bias voltage. This voltage is derived from two V_{BE} drops which are relatively constant as a function of I_C, β, and T. The actual base voltage drop is formed by

$$V_{B1,2} = 2V_{BE} + I_Q R_{B4} - I_{\dot{B}} R_{B1,2} \qquad (9.32)$$

142 INDUCTORS AND TRANSFORMERS FOR SI RF ICS

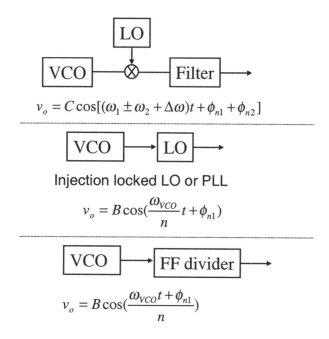

Figure 9.14. Different techniques to convert the VCO frequency: down-conversion, mode-locking, and latch-based frequency division.

Since I_B is a function of I_Q, a first order cancellation is provided by the second and third term in the above equations. Power is coupled capacitively from the VCO core to the frequency divider and output buffer circuits.

5.1   FREQUENCY DIVIDERS

Fig. 9.14 shows different possible techniques to divide the VCO oscillation frequency to a convenience frequency. First we see a traditional mixer-based approach with an LO signal driving a mixer and an on-chip filter selecting the appropriate desired harmonic. This approach is costly in terms of power as it requires several additional components. Also, the LO signal needs to be much cleaner than the VCO in terms of phase noise in order not to degrade the phase noise performance. But since the fixed LO is limited to the same on-chip LC tanks, it is likely that the LO phase noise will be at least as bad as the VCO. Hence, the output of the system suffers from more phase noise.

Another option shown next in Fig. 9.14, is to injection lock the LO to the VCO signal [Rategh and Lee, 1999]. Here, due to injection locking, the phase noise performance of the output follows the phase noise of the VCO for close-in phase noise. For proper locking, the power coupled from the VCO to the LO must exceed the intrinsic noise of the LO.

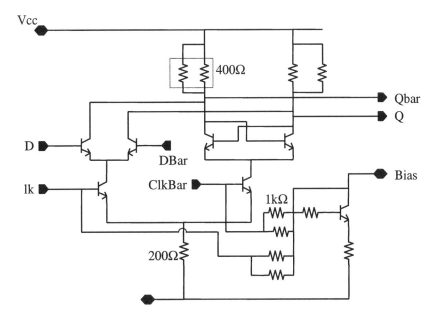

Figure 9.15. Schematic of differential latch.

The third option, and the most common, is to use latch-based frequency dividers. The advantage of using dividers is that the phase noise actually improves as we divide the frequency.

5.1.1 LATCH-BASED DIVIDERS

The differential latch of Fig. 9.15 is used at the heart of the frequency divider. The traditional master-slave topology is used to divide by two. The current consumption is 250 μA per latch and the circuit is operational up to 8 GHz. To improve headroom, the dividers are designed with emitters driving a resistor current source. A mirror biases the latch current level.

SpectreRF periodic steady state noise analysis is used to calculate the noise contribution of the frequency divider. Simulation shows that the noise contribution of the frequency divider is only significant far from the carrier.

5.2 OUTPUT BUFFERS

Figure 9.16 shows the schematic of the output buffers which are capable of delivering -12 dBm into 50 Ω. One buffer is driven by the VCO directly while another buffer is driven at the divided frequency. On-board matching must be used to improve the power gain of these devices. The buffers consume 5 mA. The design is a traditional two stage differential amplifier with input emitter follower buffers to avoid loading the output of the latches.

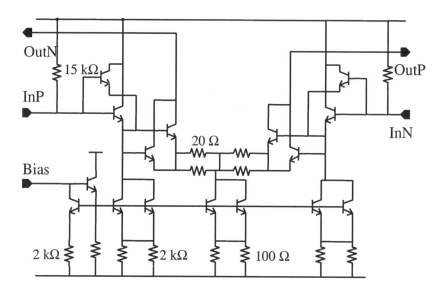

Figure 9.16. Schematic of output buffers.

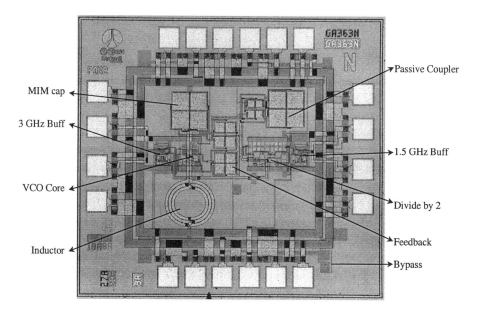

Figure 9.17. Chip-level die photo of 2.9 GHz design utilizing a circular spiral.

6. MEASUREMENTS

A chip-level layout of the final 2.9 GHz design is shown in Fig. 9.17. The 4.4 GHz layout is similar with the exception that no MIM capacitors are used

Voltage Controlled Oscillators 145

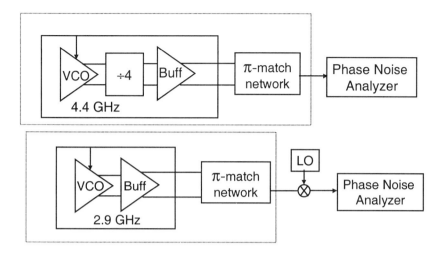

Figure 9.18. Measurement setup.

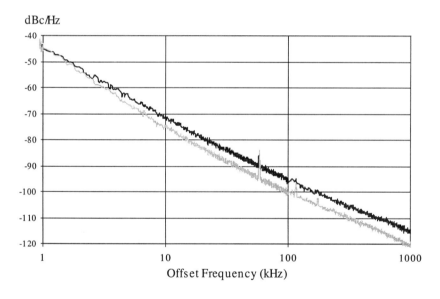

Figure 9.19. Measured phase noise of VCO.

in the tank. On-board transformers and matching networks are used to convert the signal into a single-ended 50 Ω environment for testing purposes.

Fig. 9.18 shows the test equipment setup. The chip operates on a 2.6-2.8 V supply voltage. The measured, simulated, and calculated phase noise of the 2.9 GHz and 4.4 GHz parts are shown in Fig. 9.19 and summarized in Table 9.2. Note that the 2.9 GHz oscillators were measured by down-converting the

Table 9.2. Summary of measured performance.

Center Freq.	2.9 GHz VCO	4.4 GHz VCO
Core current	3.5 mA	4.0 mA
Tuning Range	250 MHz (10%)	260 MHz (6%)
SSB Phase Noise (100 kHz offset)	-95.2 dBc/Hz	-100.2 dBc/Hz
Technology	25 GHz bipolar	
Substrate	10 Ω-cm	

carrier to near 1 GHz. This was necessary since the RDL model NTS-1000B phase noise analyzer requires a carrier below 1 GHz. Since the phase noise of the LO driving the down-converter is assumed to be better than the VCO under test, the measured phase noise is approximately the same as the phase noise of the VCO. The 4.4 GHz oscillator, though, was measured using the built-in dividers that divided the carrier by a factor of four.

The phase noise of the 2.9 GHz part is superior since the effective tank Q is higher. The oscillator was designed for 5.5 GHz but the tank resonates at 4.4 GHz due to incorrect models for the varactor. The models for the varactor were measured at 1 GHz and extrapolated. Not only is the inductor Q lower than optimal, the varactor loads the tank in the 4.4 GHz design due to the absence of linear capacitors. Measurements also confirm the 6 dB theoretical improvement in phase noise performance due to frequency division by each factor of two.

7. CONCLUSION

In this chapter we have shown that Si inductors above 3 GHz are feasible and in fact desirable. Their application, though, necessitates accurate and efficient analysis of the various loss mechanisms present in the Si IC. We have also shown that differential operation is superior to single-ended designs for fully-integrated VCOs. The phase noise of the oscillator has been calculated and verified with SpectreRF simulations. Measurements on a practical design are close to expectations.

Notes

1 It should be noted that the work of Hajimiri [Hajimiri and Lee, 1998] gives good insight into the time-varying nature of the noise and its relation to the collector current waveform.
2 This subject is treated in [Hajimiri and Lee, 1998] and is further neglected in the following analysis. The ramifications of this simplification will be noted when we compare our results with SpectreRF simulations.
3 Component values are given for the 2.9 GHz design and for the 4.5 GHz design in parentheses.

Chapter 10

DISTRIBUTED AMPLIFIERS

1. INTRODUCTION

A schematic of a distributed amplifier [Percival, 1937, Ayasli et al., 1982] is shown in Fig. 10.1. The distributed amplifier is composed of two coupled lumped transmission lines. Power is coupled from the "gate-line" to the "drain-line" through transistors M1-Mn whereas power is coupled in the reverse direction parasitically through the feedback capacitor C_{gd} of the transistors. Since for a practical transistor $s_{21} \gg s_{12}$, the behavior of the amplifier is dominated by the "forward-direction" coupling of power from the gate line to the drain line. As evident from the figure, the parasitics of the transistor form an integral part of the amplifier, completing the transmission lines. In this way, the parasitics of the amplifier do not limit the gain-bandwidth product of the amplifier as in a traditional amplifier, allowing large bandwidths to be achieved. If care is exercised in equalizing the phase velocity on the gate and drain line, then power

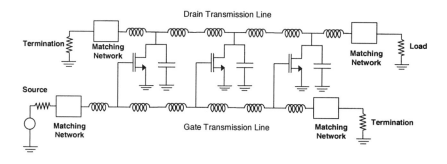

Figure 10.1. A three-stage distributed amplifier using CMOS n-FETs.

interferes constructively and the overall gain of the amplifier is enhanced by the addition of each stage.

The number of stages that may be used is limited by the inherent losses on the drain and gate line. Thus, the additional attenuation incurred by the addition of a stage may outweigh the benefits of gain and thus there is an optimum number of stages for gain. This will always occur since the gain is a polynomial of the number of stages, whereas the transmission line attenuation is a decaying exponential function of the length of the transmission line.

In Section 2. we will review the image parameter method as it relates to designing a distributed amplifier. Specially, the inherent losses of the transistor and microstrips will be taken into account to find the line impedance, propagation and attenuation constants. These results will be used in Section 3. where we derive the gain of a distributed amplifier.

2. IMAGE PARAMETER METHOD

The image parameter method may be applied to the distributed amplifier since it consists of a cascade of identical two-port networks forming an artificial transmission line. The image impedance Z_i for a reciprocal symmetric two-port is defined as the impedance looking into port 1 or 2 of the two-port when the other terminal is also terminated in Z_i. This impedance is given by [Pozar, 1997]

$$Z_i = \sqrt{\frac{B}{C}} \tag{10.1}$$

where the $ABCD$ matrix of the two-port is given by

$$\begin{pmatrix} A & B \\ C & D \end{pmatrix} \tag{10.2}$$

where $A = D$ by symmetry. The propagation constant for the current and voltage is given by

$$e^{-\gamma} = \sqrt{AD} - \sqrt{BC} \tag{10.3}$$

Thus, wave propagation occurs if γ has an imaginary component. For the T-network shown in Fig. 10.2, the image impedance and propagation factors are given by

$$Z_i = \sqrt{Z_1 Z_2}\sqrt{1 + \frac{Z_1}{4Z_2}} \tag{10.4}$$

$$e^{\gamma} = 1 + \frac{Z_1}{2Z_2} + \sqrt{\frac{Z_1}{Z_2} + \frac{Z_1^2}{4Z_2^2}} \tag{10.5}$$

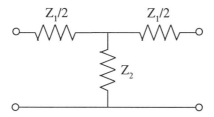

Figure 10.2. T-Section Network.

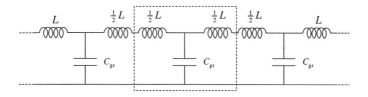

Figure 10.3. Lossless cascade of T-sections forming an artificial transmission line.

2.1 LOSSLESS LUMPED TRANSMISSION LINE

For a lossless cascade of T-sections such as that shown in Fig. 10.3, the propagation factor and image impedance are given by [Pozar, 1997]

$$Z_i = \sqrt{\frac{L}{C}\left(1 - \left(\frac{\omega}{\omega_c}\right)^2\right)} \qquad (10.6)$$

$$e^\gamma = 1 - \frac{2\omega^2}{\omega_c^2} + \frac{2\omega}{\omega_c}\sqrt{\left(\frac{\omega}{\omega_c}\right)^2 - 1} \qquad (10.7)$$

$$\omega_c = \frac{2}{\sqrt{LC}} \qquad (10.8)$$

Thus, for $\omega < \omega_c$ the lossless structure has a real image impedance and a purely imaginary propagation factor. Beyond the cutoff frequency ω_c, the propagation factor has a real component that increases without bound. These results are consistent with the distributed transmission line since $\omega_c \to \infty$ for such a structure.

2.2 LOSSY LUMPED TRANSMISSION LINE

Figures 10.4 and 10.5 show the lossy "gate" and "drain" transmission line. The losses are due to the lossy microstrips or spiral inductors as well as the losses of the input/output impedance looking into the gate/drain of the FET. A

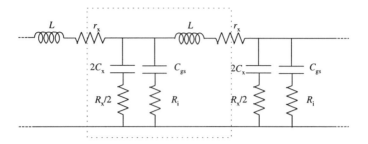

Figure 10.4. Lossy "gate" transmission line.

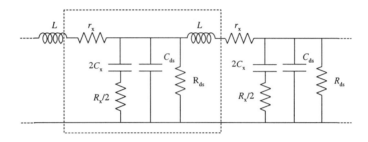

Figure 10.5. Lossy "drain" transmission line.

series gate resistance models the FET losses. The inductor losses are divided into two components, one part r_x modeling the metallization losses in series with the inductor and one component R_x modeling the substrate losses. For an insulating substrate, the conductive substrate losses are negligibly small and so $R_x = 0$. If dielectric losses are present, it is more appropriate to place R_x in shunt rather than in series with the substrate capacitance.

In reality these loss resistors are frequency-dependent owing to skin and proximity effects and displacement current in the substrate, but for simplicity we will assume a constant value. If the substrate is sufficiently conductive and near the microstrip metal layers, induced eddy currents lead to an additional loss mechanism. This is the case for highly conductive substrates used in epi CMOS processes. Again, we neglect this loss in the following analysis.

Using the notation of the previous section, the T-section impedance and admittance are

$$Z_{1g} = j\omega L_g + r_{xg} \qquad (10.9)$$

$$Y_{2g} = \frac{j\omega C_{gs}\left(1 + 2\frac{C_{xg}}{C_{gs}}\right) - \omega^2 C_{gs}\left(\frac{1}{\omega_{sg}} + \frac{C_{xg}}{\omega_g C_{gs}}\right)}{\left(1 + j\frac{\omega}{\omega_g}\right)\left(1 + j\frac{\omega}{\omega_{xg}}\right)} \qquad (10.10)$$

where
$$\omega_g = \frac{1}{C_{gs}R_i} \tag{10.11}$$

and
$$\omega_{xg} = \frac{1}{C_{gs}R_{xg}} \tag{10.12}$$

Similarly, for the lossy drain line, the T-section impedance and admittance are
$$Z_{1d} = j\omega L_d + r_{xd} \tag{10.13}$$

$$Y_{2g} = \frac{G_{ds}\left(1 - \frac{\omega^2}{\omega_{xd}\omega_{ds}}\right) + j\omega\left(\frac{2G_{xd}}{\omega_{xd}} + \frac{G_{ds}}{\omega_{ds}} + \frac{G_{ds}}{\omega_{xd}}\right)}{\left(1 + j\frac{\omega}{\omega_{xd}}\right)} \tag{10.14}$$

where
$$\omega_d = \frac{1}{C_{ds}R_{ds}} \tag{10.15}$$

and
$$\omega_{xd} = \frac{1}{C_{xd}R_{xd}} \tag{10.16}$$

Using (10.9)(10.10)(10.13) and (10.14), the image impedance and propagation factor can be calculated using (10.4) and (10.5). These expressions are plotted in Figures 10.6, 10.7, and 10.8. The plots are generated using the following gate and drain parameters

$$\begin{array}{ll} L_g = 1.85 \text{ nH} & r_{xg} = \omega_{Lg}L_g \\ C_{gs} = 310 \text{ fF} & R_i = 7.25\Omega \\ C_{xg} = 50 \text{ fF} & R_{xg} = \frac{1}{\omega_{xg}C_{xg}} \\ L_d = 2.04 \text{ nH} & r_{xd} = \omega_{Ld}L_d \\ C_{ds} = 96 \text{ fF} & R_{ds} = 385\Omega \\ C_{xd} = 50 \text{ fF} & R_{xd} = \frac{1}{\omega_{xd}C_{xd}} \end{array} \tag{10.17}$$

The FET parameters come from a hypothetical transistor model, whereas the drain and gate inductor parasitics are shown parameterized as a function of the cutoff frequencies. In Fig. 10.6 we also include the lossless image impedance given by (10.6) for comparison.

As evident from the figures, even for lossless passive devices the losses in the intrinsic FET warrant careful analysis. Additionally, the frequency dependence of the image impedance can potentially destroy the broadband operation of the device if it is not matched correctly.

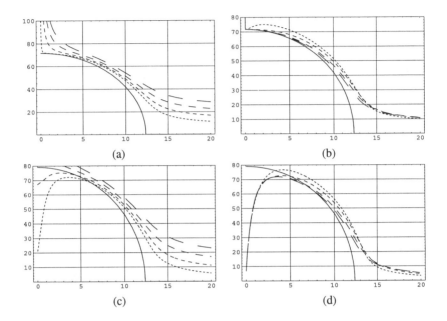

Figure 10.6. Gate (curves (a) and (b)) and drain (curves (c) and (d)) image impedance as a function of inductor and capacitor loss cutoff frequency. In (a), starting from the bottom curve, the gate inductor loss cutoff frequency is 0 (lossless), .1 GHz, 1 GHz, 2 GHz, and 3 GHz while the capacitance loss cutoff frequency is held constant at 9 GHz. In (b), starting from the bottom curve, the gate capacitor loss cutoff frequency is ∞ (lossless), 15 GHz, 10 GHz, 5 GHz, and 1 GHz while the inductor loss cutoff frequency is held constant at .01 GHz. In (c), starting from the bottom curve, the drain inductor loss cutoff frequency is 0 (lossless), .1 GHz, 1 GHz, 2 GHz, and 3 GHz while the capacitance loss cutoff frequency is held constant at 9 GHz. In (d), starting from the bottom curve, the drain capacitor loss cutoff frequency is ∞ (lossless), 15 GHz, 10 GHz, 5 GHz, and 1 GHz while the inductor loss cutoff frequency is held constant at .01 GHz.

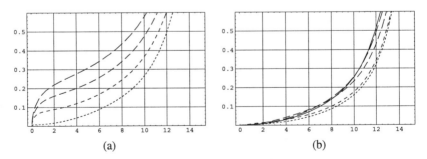

Figure 10.7. Gate line attenuation as a function of inductor and capacitor loss cutoff frequency. In (a), starting from the bottom curve, the gate inductor loss cutoff frequency is 0.1 GHz, 1 GHz, 2 GHz, and 3 GHz while the capacitance loss cutoff frequency is held constant at 9 GHz. In (b), starting from the bottom curve, the gate capacitor loss cutoff frequency is 15 GHz, 10 GHz, 5 GHz, 1 GHz, and 0.1 GHz while the inductor loss cutoff frequency is held constant at .01 GHz.

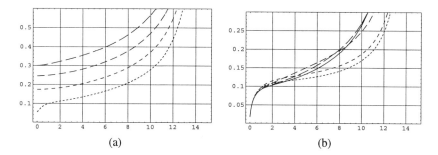

Figure 10.8. Drain line attenuation as a function of inductor and capacitor loss cutoff frequency. In (a), starting from the bottom curve, the drain inductor loss cutoff frequency is 0.1 GHz, 1 GHz, 2 GHz, and 3 GHz while the capacitance loss cutoff frequency is held constant at 9 GHz. In (b), starting from the bottom curve, the drain capacitor loss cutoff frequency is 15 GHz, 10 GHz, 5 GHz, 1 GHz, and 0.1 GHz while the inductor loss cutoff frequency is held constant at .01 GHz.

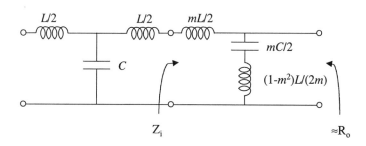

Figure 10.9. Bisected-π m-derived matching network.

2.3 IMAGE IMPEDANCE MATCHING

To achieve an impedance match over a broad range, the load and source impedance must be transformed into the line image impedance. Otherwise, the gain response will not be flat as a function of frequency.

The bisected-π m-derived section shown in Fig. 10.9, serves this purpose well [Pozar, 1997]. In Fig. 10.10 we plot the impedance looking into the gate and drain line when transformed by the m-derived section. As evident from the figure, the impedance is approximately constant over a broad range of frequencies, a big improvement over the frequency variation of the image impedance in Fig. 10.6. The m-derived impedance matching network can also be used to match directly to $Z_0 = 50\ \Omega$.

The performance of the m-derived section shown in Fig. 10.10 includes the lossy transmission line as well as the ideal transmission line. The lossy case performance is not as good as the lossless but it can be flattened with further optimization.

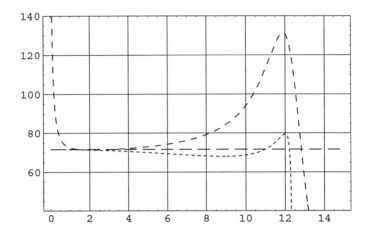

Figure 10.10. Gate line input impedance using bisected-π m-derived matching network. Bottom curve shows the performance of a lossless transmission line. Top curve shows performance with lossy matching network. The flat line is the ideal gate impedance.

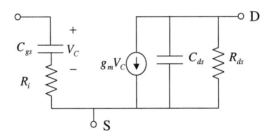

Figure 10.11. Simple FET model used for analysis.

3. DISTRIBUTED AMPLIFIER GAIN

Given the background of Section 2., we are now in a good position to derive the gain of the distributed amplifier. The following derivation closely follows the work of [Beyer et al., 1984] but deviates in that we work directly with the complete expressions for the transmission line parameters as opposed to the approximate expressions developed by [Beyer et al., 1984].

3.1 EXPRESSION FOR GAIN

Using the simplified FET model shown in Fig. 10.11, the total output current of an n-stage distributed amplifier shown in Fig. 10.1 can be written as

$$I_o = \frac{1}{2} g_m e^{-\gamma_d/2} \sum_{k=1}^{n} V_{ck} e^{-(n-k)\gamma_d} \qquad (10.18)$$

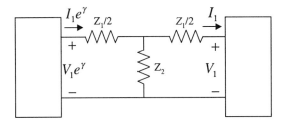

Figure 10.12. Derivation of gate-voltage at the center of the T-section.

where γ_d is the propagation delay of the drain line given by (10.5) and V_{ck} is the voltage drop across the kth stage input capacitor. Note that we have assumed constructive interference of the FET currents since in a practical design the phase velocity of the drain line is matched to the gate line by adding additional shunt drain capacitance to satisfy

$$\beta = \Im\{\gamma_g\} = \Im\{\gamma_d(C_p)\} \tag{10.19}$$

In the low loss case this is equivalent to

$$\frac{2}{\sqrt{L_g(C_{gs} + 2C_{xg})}} = \frac{2}{\sqrt{L_d(C_{ds} + 2C_{xd} + C_p)}} \tag{10.20}$$

The above relation can be used as a starting point solution in a non-linear iteration loop to solve (10.19). (10.19) represents a continuum of equations since it varies as a function of ω, and therefore it can be satisfied in the least squared sense over the range of interest by minimizing

$$\int_0^{\omega_c} (\Im(\gamma_g) - \Im(\gamma_d))^2 \, d\omega \tag{10.21}$$

The voltage across the kth stage input capacitor is given by

$$V_{ck} = \frac{V_i \delta \exp\left(\frac{-(2k-1)\gamma_g}{2} - j\tan\frac{\omega}{\omega_g}\right)}{\sqrt{1 + \left(\frac{\omega}{\omega_g}\right)^2}} \tag{10.22}$$

The origin of the exponential term is due to the finite propagation velocity along the gate line as well as the frequency-dependent voltage division between the input capacitance and the resistance. The term δ is the voltage at the center of the T-section. This can be derived with the aid of Fig. 10.12. Note that

$$I_1 e^{\gamma} = Y_2 V_c + I_1 \tag{10.23}$$

and
$$\frac{V_c - V_1}{Z_1/2} = I_1 \tag{10.24}$$

Using the above relations to eliminate I_1 one obtains

$$\frac{V_c}{V_1} = \frac{e^\gamma - 1}{e^\gamma - \left(1 + \frac{Z_1}{2Z_2}\right)} \tag{10.25}$$

Making use of (10.5) one obtains

$$\delta \equiv \frac{V_c}{V_1} = \frac{\frac{Z_1}{2Z_2} + \sqrt{\frac{Z_1}{Z_2} + \left(\frac{Z_1}{2Z_2}\right)^2}}{\sqrt{\frac{Z_1}{Z_2} + \left(\frac{Z_1}{2Z_2}\right)^2}} \tag{10.26}$$

Using (10.22) in (10.18) one obtains

$$I_o = \frac{g_m V_i \delta \sinh\left[\frac{n}{2}(\alpha_d - \alpha_g)\right] \exp\left[\frac{-n(\alpha_d + \alpha_g)}{2}\right] \exp\left[-jn\phi - j\tan^{-1}\frac{\omega}{\omega_g}\right]}{2\sqrt{1 + \left(\frac{\omega}{\omega_g}\right)^2} \sinh\left[\frac{1}{2}(\alpha_d - \alpha_g)\right]} \tag{10.27}$$

To find the power gain, we use the following expressions

$$P_o = \frac{1}{2}|I_o|^2 \Re\left[Z'_{id}\right] \tag{10.28}$$

$$P_i = \frac{|V_i|^2}{2|Z_{ig}|^2} \Re\left[Z'_{ig}\right] \tag{10.29}$$

Thus, under matched conditions we obtain

$$G = \frac{\Re\left[Z'_{id}\right]}{\Re\left[Z'_{ig}\right]}|Z_{ig}|^2 \frac{g_m^2 |\delta|^2 \sinh^2\left[\frac{n}{2}(\alpha_d - \alpha_g)\right] \exp\left[-n(\alpha_d + \alpha_g)\right]}{4\left[1 + \left(\frac{\omega}{\omega_g}\right)^2\right] \sinh^2\left[\frac{1}{2}(\alpha_d - \alpha_g)\right]} \tag{10.30}$$

where the primed impedances are the transformed line image impedance (using, for instance, an m-derived matching section).

3.2 DESIGN TRADEOFFS

The above expression (10.30) is the gain as a function of the drain/gate transmission line parameters, assuming fixed FET parameters. In reality, the FET width and length may be scaled as well. But typically for highest frequency response performance the minimum allowed length is used. The width can be selected according to power handling capability. The physical layout of the device should minimize the input gate resistance.

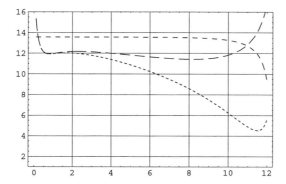

Figure 10.13. Gain as a function of frequency using Beyer's ideal expression (top curve) and the gain calculated in this work with an m-derived matching network (flat increasing curve) and without a matching network in place (rapidly decaying curve).

The parameters under direct control of the designer are thus the number of stages, n, and the gate and drain inductance, L_g and L_d. Hence for a given number of stages, there are only two parameters to vary. To achieve a good match to 50 Ω, one generally chooses L_g and L_d to set the gate and drain impedances as close to 50 Ω as possible, limiting the range of values for these components tremendously. There seems to be only a single parameter n in the design of a distributed amplifier.

In practice one can trade gain for bandwidth by increasing the gate and drain cutoff frequencies through adding a series capacitor to the gate of the input stages. The design tradeoff must be evaluated carefully, though, since a lower gain with more bandwidth can also be achieved by simply reducing n. In Fig. 10.13 we plot the value of gain versus frequency for the FET transistor ($n = 4$) using the gate and drain line parameters given in (10.17). Also plotted is the expression for gain derived by [Beyer et al., 1984]. The deviation of ideal flat behavior is actually mostly due to the non-constant image impedance. This is why the m-derived matching network is necessary. In Fig. 10.13 we also show the gain for a matched design. Note that the matching network improves the gain flatness and is a critical part of the distributed amplifier.

Also evident from (10.30) is the strong dependence on the gate and drain attenuation factors. As discussed before, the attenuation factors in fact set the optimum number of stages for gain. Naively, the gain should increase as we add stages, as shown by the approximate expression of [Beyer et al., 1984] where n^2 dependence is shown. But eventually the attenuation on the gate line will drive the input voltage to negligibly small values and adding further stages will simply increase the length and hence attenuation on the drain line without contributing to the output current. This is shown in Fig. 10.14, where the gain

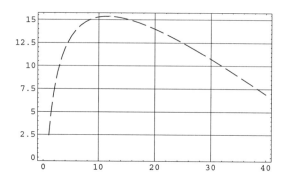

Figure 10.14. Gain versus the number of stages evaluated at low frequency (2 GHz).

at 2 GHz is plotted as a function of n (a low frequency is selected so that the effects of gain roll-off will not come into play). The optimum number of stages may be calculated from evaluating the derivative of (10.30) with respect to n

$$n_{opt} = \frac{\ln \frac{\alpha_d}{\alpha_g}}{\alpha_d - \alpha_g} \qquad (10.31)$$

The above expression is frequency dependent and may be evaluated to optimize the low frequency gain. Otherwise, we may use effective average values of the propagation loss factors

$$\overline{\alpha} = \frac{1}{\omega_c} \int_0^{\omega_c} |\alpha(\omega)| d\omega \qquad (10.32)$$

To extend the frequency response, coupled inductors in the form of T-coils may be employed [Vendelin and A. M. Pavio, 1990]. This topology has a cutoff frequency $\sqrt{2}$ times higher than the topology analyzed in this paper.

3.3 ACTIVELY LOADED GATE LINE

The above analysis identifies the attenuation loss factors as the main design constraint in achieving large values of gain by limiting the number of practical stages one can employ. In a modern CMOS process, these factors are the dominant limiting factors since high Q passive devices are difficult to implement due to the conductive substrate.

To boost the AC gain at the expense of DC power, the circuit topology of Fig. 10.15 may be used. For stability, the negative resistance must be chosen so that the total real impedance has a positive real part at all frequencies of interest. In practice, it will be difficult to achieve a constant negative resistance over a broad range of frequencies. This is not a big limitation since the frequency dependence of the $-R$ stage can help overcome the frequency dependence in α.

Distributed Amplifiers 161

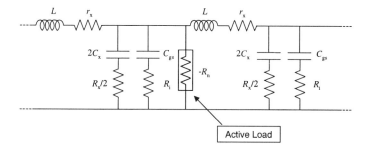

Figure 10.15. Active load proposal to reduce gate or drain line losses.

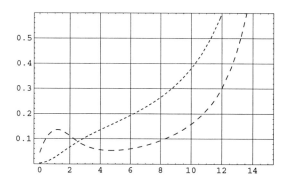

Figure 10.16. Actively loaded gate attenuation constant in action. Propagation constant remains imaginary (not shown).

The design equations for the actively loaded gate transmission line are the following along with (10.4) and (10.5)

$$Z_1 = j\omega L + r_x \qquad (10.33)$$

$$Y_{2g} = \frac{G_n + \frac{\omega^2}{\omega_g \omega_x}(G_n - G_x - G_i) + j\omega\left(\frac{2G_x - G_n}{\omega_x} + \frac{G_i - G_n}{\omega_g}\right)}{\left(1 + j\frac{\omega}{\omega_x}\right)\left(1 + j\frac{\omega}{\omega_g}\right)} \qquad (10.34)$$

In Fig. 10.16 we plot the gate loss with and without the $-R$ stage. The reduction in the attenuation loss is clear.

Chapter 11

CONCLUSION

The focus of this book has been the analysis and applications of passive devices in Si RF and microwave ICs. As we have seen, these devices play a critical role in modern monolithic transceivers and high speed analog circuits and their analysis warrants careful attention. With clock frequencies exceeding 1 GHz, the careful analysis of inductive effects in digital circuits is increasingly a concern as well.

Part I of the book presented efficient and accurate analysis techniques which give physical insight into the device operation at high frequency. The partial element equivalent circuits (PEEC) concept has been used to discretize the constituent conductors of the device. Quasi-static 2D and 3D Green functions were employed to account for the response of the conductive substrate. In this way, both eddy currents in the metallization as well as in the bulk substrate were taken into account.

Measurements on a highly conductive substrate confirm the validity of the approach. *ASITIC* simulations predict the electrical parameters of single layer and multi-layer inductors from 200 MHz–12 GHz. In particular, the inductance, the loss, and self-resonant frequency were predicted to within experimental error tolerances. Previous measurements on a moderately resistive substrate using single layer and multi-layer polygon inductors, planar and non-planar transformers, baluns, and coupled inductors also validated the techniques for a wide range of structures.

Part II of the book examined critical RF and microwave circuit building blocks which depend critically on passive devices. The design of the voltage-controlled oscillator was presented with particular attention to the passive devices. We demonstrated the feasibility of high frequency VCOs using standard Si technology fabrication techniques. In particular, the passive devices and layout parasitics are designed and optimized with *ASITIC*. We also presented an integrated Si distributed amplifier for future broadband communications.

1. FUTURE RESEARCH
1.1 LARGER PROBLEMS

One obvious extension of the present work is to tackle larger problems. The techniques of this book may be applied to moderately large problems, such as a small RF chip with several inductors, capacitors, and high frequency metallization traces. Larger problems, though, require more efficient numerical techniques. For instance, one may wish to analyze an entire RF/baseband chip, including the bond pads, the metallization, the substrate coupling, and the passive devices. These problems involve millions of elements and a full 3D analysis is prohibitively expensive. On the other hand, judicious numerical and physical approximation techniques can break the problem into many sub-problems. The matrix computations will no doubt involve efficient iterative solvers as opposed to the direct solvers used in the present work. Sparsification of the matrices or operators can help in this regard.

1.2 DIGITAL CIRCUITS

Microprocessors today are clocked at frequencies above 1 GHz and this trend will continue into the future. The main bottleneck will be the interconnection and not the transistor performance. To overcome this bottleneck, digital designers will need to incorporate models for the inductance and magnetic coupling of metal traces as well as the capacitive losses and coupling. Substrate coupling and loss will also be an issue. The challenge with digital circuits is of course the large volume and count of interconnection that need to be analyzed. Since digital circuits are more fault tolerant, perhaps tradeoffs in the accuracy of the numerical techniques can accelerate the solution. The coupling from noisy digital circuits in the analog and RF signal paths is also an interesting extension of the techniques of this research. For instance, digital gates can be represented by equivalent capacitors injecting pseudo-random noise into the substrate. This noise will be injected into sensitive analog nodes and will lower the signal-to-noise ratio and ultimately the bit error rates in the system.

1.3 MEMS TECHNOLOGY

The modern IC process now allows the construction of micro-electro-mechanical structures (MEMS). These devices open up a wealth of possibilities for designing passive devices. Simulation and analysis of these structures presents many new challenges as Maxwell's equations must be solved simultaneously with mechanical and fluid dynamics equations, often involving non-linear terms. Such mixed-domain simulations require careful numerical techniques.

Conclusion 165

As an example of an electro-mechanical passive device, consider a spiral inductor constructed on metal layers but suspended from the Si substrate and hinged, allowing it free motion. A patterned shield plate is placed under the inductor, forming a resonant tank. The self-resonant frequency of this tank can be tuned by applying an electrostatic potential to the inductor, rotating the spiral and adjusting the capacitance. In addition to the ability to tune the center frequency of this structure, this device is likely to have fewer losses as the device is somewhat isolated from the lossy substrate due to its vertical placement. The magnetic fields are now parallel as opposed to perpendicular to the substrate and hence eddy currents are curtailed. Displacement current is also reduced as the structure is shielded and isolated from the substrate. Optimization of this structure can proceed by minimizing the skin and proximity effects in the conductors using the techniques presented in this research. Modeling the mechanical motion of the hinge, though, requires techniques beyond the scope of this research.

Appendix A
Distributed Capacitance

Consider a distributed lossy capacitor of length ℓ, represented schematically in Fig. A.1. Let C_T denote the total capacitance and R_T the total resistance of the structure. In this appendix we will show that in the limit of an electrically short open-line, the equivalent circuit for such a structure is simply the total capacitance in series with $R_T/3$.

Define the capacitance and resistance per unit length

$$C = \frac{C_T}{\ell} \quad \text{(A.1)}$$

$$R = \frac{R_T}{\ell} \quad \text{(A.2)}$$

For a short section of length δx of the device, shown in Fig. A.2, we have

$$v(x,t) = (R\delta x)i(x,t) + v(x+\delta x, t) \quad \text{(A.3)}$$

$$i(x,t) = i(x+\delta x, t) + (C\delta x)\frac{\partial V(x+\delta x, t)}{\partial t} \quad \text{(A.4)}$$

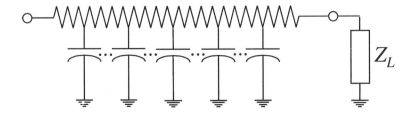

Figure A.1. Schematic of a distributed lossy capacitor terminated in an arbitrary impedance.

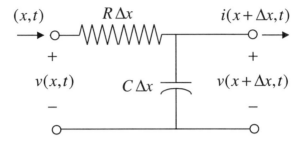

Figure A.2. A short segment of the distributed lossy capacitor.

Letting $\delta x \to 0$

$$\frac{\partial V}{\partial x} = -Ri \tag{A.5}$$

$$\frac{\partial i}{\partial x} = -C\frac{\partial V(x + \delta x, t)}{\partial t} \tag{A.6}$$

For the time-harmonic case, we have the following set of coupled ordinary differential equations

$$\frac{dV}{dx} = -Ri \tag{A.7}$$

$$\frac{di}{dx} = -j\omega CV \tag{A.8}$$

The equations are decoupled by taking the derivative of the above equations

$$\frac{d^2V}{dx^2} = -R\frac{di}{dx} = j\omega RCV \tag{A.9}$$

$$\frac{d^2i}{dx} = -j\omega C\frac{dV}{dx} = j\omega RCi \tag{A.10}$$

The above boundary value problem can be solved once the following boundary conditions are imposed

$$V(0) = V_i \tag{A.11}$$

$$V(\ell) = i(\ell)Z_L \tag{A.12}$$

where Z_L is the load impedance terminating the lossy capacitor. Consider the eigenfunction solution $V(x) = e^{\alpha x}$. Since $V''(x) = \alpha^2 V(x)$ we require that

$$\alpha^2 = \frac{i\omega R_T C_T}{\ell^2} \tag{A.13}$$

Appendix A: Distributed Capacitance   169

or
$$\alpha = \sqrt{\frac{j\omega R_T C_T}{\ell^2}} = \sqrt{\frac{\omega}{\omega_x}}\frac{\sqrt{j}}{\ell} \tag{A.14}$$

where $\omega_x \equiv \frac{1}{R_T C_T}$ so that

$$\alpha = \pm\frac{\sqrt{2}}{2}\frac{1+j}{\ell}\sqrt{\frac{\omega}{\omega_x}} \tag{A.15}$$

Note that the propagation and attenuation constant are equal, $\Im(\alpha) = \Re(\alpha)$. The general solution is thus

$$v(x) = Ae^{\alpha x} + Be^{-\alpha x} \tag{A.16}$$
$$i(x) = Me^{\alpha x} + Ne^{-\alpha x} \tag{A.17}$$

The unknown coefficients are found by imposing the boundary conditions. The input impedance is therefore

$$\begin{aligned}Z_{in} &= \frac{v(0)}{i(0)} = \frac{V_i}{M+N} \\ &= \left(\frac{-\alpha}{\omega C}\right)\frac{-jCZ_L\omega + \alpha\tanh(\alpha\ell)}{j\alpha + CZ_L\omega\tanh(\alpha\ell)} \end{aligned} \tag{A.18}$$

Consider the limit of a short line at zero frequency such that $\alpha\ell \to 0$ and $\omega \to 0$. Then we have $Z_{in} = Z_L$ as expected. Similarly, consider a shorted line such that $Z_L \to 0$. Then

$$Z_{in} = \frac{\alpha}{j\omega C}\tanh(\alpha\ell) \tag{A.19}$$

If we consider now the above in the limit of a short line, we have $Z_{in} \approx R_T$ as expected.

The limiting case of interest is that of an open line. Taking the limit $Z_L \to \infty$ we have

$$Z_{in} = \frac{\alpha}{j\omega C}\coth(\alpha\ell) \tag{A.20}$$

Expanding the hyperbolic cotangent function in a Taylor series we have

$$Z_{in} \approx \frac{R_T}{3} + \frac{1}{j\omega C_T} \tag{A.21}$$

This is what we set out to prove.

References

[Abramowitz and Stegun, 1972] Abramowitz, M. and Stegun, I. A. (1972). *Handbook of Mathematical Functions with Formulas, Graphs, and Mathematical Tables*. John Wiley and Sons, Inc., New York.

[Ashby et al., 1996] Ashby, K. B., Koullias, I. A., Finley, W. C., Bastek, J. J., and Moinian, S. (1996). High q inductors for wireless applications in a complementary silicon bipolar process. *IEEE Journal of Solid-State Circuits*, 31:4–9.

[Ayasli et al., 1982] Ayasli, Y., Mozzi, R. L., Vorhaus, J. L., Reynolds, L. D., and Pucel, R. A. (1982). A monolithic gaas 1-13 ghz traveling-wave amplifier. *IEEE Trans. Microwave Theory Tech.*, MTT-30:976–981.

[Beyer et al., 1984] Beyer, J. B., Prasad, S. N., Becker, R. C., Nordman, J. E., and Hohenwarter, G. K. (1984). Mesfet distributed amplifier design guidelines. *IEEE Trans. Microwave Theory Tech.*, MTT-32:268–275.

[Boulouard and Le Rouzic, 1989] Boulouard, A. and Le Rouzic, M. (1989). Analysis of rectangular spiral transformers for mmic applications. *IEEE Transactions on Microwave Theory and Techniques*, 37:1257–60.

[Burghartz et al., 1996a] Burghartz, J. N., Jenkins, K. A., and Soyuer, M. (1996a). Multilevel-spiral inductors using vlsi interconnect technology. *IEEE Electron Device Letters*, 17:428–30.

[Burghartz et al., 1997] Burghartz, J. N., Ruehli, A. E., Jenkins, K. A., Soyuer, M., and Nguyen-Ngoc, D. (1997). Novel substrate contact structure for high-q silicon-integrated spiral inductors. In *International Electron Devices Meeting*, pages 55–8.

[Burghartz et al., 1996b] Burghartz, J. N., Soyuer, M., and Jenkins, K. A. (1996b). Integrated rf and microwave components in bicmos technology. *IEEE Transactions on Electron Devices*, 43:1559–70.

[Burghartz et al., 1995] Burghartz, J. N., Soyuer, M., Jenkins, K. A., and Hulvey, M. D. (1995). High-q inductors in standard silicon interconnect technology and its application to an integrated rf power amplifier. In *International Electron Devices Meeting*, pages 1015–8.

[Cahana, 1983] Cahana, D. (1983). A new transmission line approach for designing spiral microstrip inductors for microwave integrated circuits. In *IEEE MTT-S International Microwave Symposium Digest*, pages 245–7.

[Carlin and Giordano, 1964] Carlin, H. J. and Giordano, A. B. (1964). *Network theory; an introduction to reciprocal and non-reciprocal circuits*. Prentice-Hall, Englewood Cliffs, N.J.

[Chang et al., 1993] Chang, J. Y.-C., Abidi, A. A., and Gaitan, M. (1993). Large suspended inductors on silicon and their use in a 2 μm cmos rf amplifier. *IEEE Electron Device Letters*, 14:246–8.

[Collin, 1990] Collin, R. E. (1990). *Field theory of guided waves*. IEEE Press, New York, 2nd edition.

[Craninckx and Steyaert, 1995] Craninckx, J. and Steyaert, M. S. J. (1995). A 1.8-ghz cmos low-phase-noise voltage-controlled oscillator with prescaler. *IEEE Journal of Solid-State Circuits*, 30:1474–82.

[Craninckx and Steyaert, 1997] Craninckx, J. and Steyaert, M. S. J. (1997). A 1.8-ghz low-phase-noise cmos vco using optimized hollow spiral inductors. *IEEE J. Solid-State Circuits*, 32:736–744.

[Danesh et al., 1998] Danesh, M., Long, J. R., Hadaway, R. A., and Harame, D. L. (1998). A q-factor enhancement technique for mmic inductors. In *IEEE MTT-S International Microwave Symposium Digest*, pages 183–6.

[Dec and Suyama, 1997] Dec, A. and Suyama, K. (1997). Micromachined varactor with wide tuning range. *Electronics Letters*, 33:922–4.

[Dekker et al., 1997] Dekker, R., Baltus, P., van Deurzen, M., Einden, W., Maas, H., and Wagemans, A. (1997). An ultra low-power rf bipolar technology on glass. In *International Electron Devices Meeting*, pages 921–3.

[Demir et al., 1998] Demir, A., Mehrotra, A., and Roychowdhury, J. (1998). Phase noise in oscillators: a unifying theory and numerical methods for characterisation. In *Design and Automation Conference*, pages 26–31.

[Desoer and Kuh, 1969] Desoer, C. A. and Kuh, E. S. (1969). *Basic circuit theory*. McGraw-Hill, New York.

[Dongarra and Demmel, 1991] Dongarra, J. and Demmel, J. (1991). Lapack: a portable high-performance numerical library for linear algebra. *Supercomputer*, 8:33–8.

[Erzgraber et al., 1998] Erzgraber, H. B., Grabolla, T., Richter, H. H., Schley, P., and Wolff, A. (1998). A novel buried oxide isolation for monolithic rf inductors on silicon. In *International Electron Devices Meeting*, pages 535–9.

[et al., 1983] et al., R. P. (1983). *Quadpack: A Subroutine Package for Automatic Integration*. Springer-Verlag, Berlin; New York.

[Eynde et al., 2000] Eynde, F. O., Craninckx, J., and Goetschalckx, P. (2000). A fully-integrated zero-if dect transceiver. In *IEEE International Solid-State Circuits Conference*, pages 138–9,450.

[Feynman et al., 1963] Feynman, R. P., Leighton, R. B., and Sands, M. L. (1963). *The Feynman lectures on physics volume 2*. Addison-Wesley Pub. Co, Reading, Mass.

[Frigo and Johnson, 1998] Frigo, M. and Johnson, S. G. (1998). Fftw: an adaptive software architecture for the fft. In *IEEE International Conference on Acoustics, Speech and Signal Processing*, pages 1381–4.

[Geen et al., 1989] Geen, M. W., Green, G. J., Arnold, R. G., Jenkins, J. A., and Jansen, R. H. (1989). Miniature multilayer spiral inductors for gaas mmics. In *11th Annual GaAs IC Symposium*, pages 303–6.

[Gharpurey and Meyer, 1995] Gharpurey, R. and Meyer, R. G. (1995). Analysis and simulation of substrate coupling in integrated circuits. *International Journal of Circuit Theory and Applications*, 23:381–394.

[Gharpurey and Meyer, 1996] Gharpurey, R. and Meyer, R. G. (1996). Modeling and analysis of substrate coupling in integrated circuits. *IEEE Journal of Solid-State Circuits*, 31(3):344–53.

[Gray and Meyer, 1993] Gray, P. R. and Meyer, R. G. (1993). *Analysis and design of analog integrated circuits*. Wiley, New York, 3rd edition.

[Gray and Meyer, 1995] Gray, P. R. and Meyer, R. G. (1995). Future directions in silicon ics for rf personal communications. In *IEEE International Solid-State Circuits Conference*, pages 83–90.

[Greengard and Rokhlin, 1997] Greengard, L. and Rokhlin, V. (1997). A fast algorithm for particle simulations. *Journal of Computational Physics*, 135:280–92.

[Greenhouse, 1974] Greenhouse, H. M. (1974). Design of planar rectangular microelectronic inductors. *IEEE Trans. Parts, Hybrids and Packaging*, PHP-10:101–9.

[Grover, 1946] Grover, F. W. (1946). *Inductance Calculations*. Van Nostrand, Princeton, N.J.

[Hajimiri and Lee, 1998] Hajimiri, A. and Lee, T. H. (1998). A general theory of phase noise in electrical oscillators. *IEEE Journal of Solid-State Circuits*, 33:179–94.

[Hasegawa et al., 1971] Hasegawa, H., Furukawa, M., and Yanai, H. (1971). Properties of microstrip line on si-sio_2 system. *IEEE Transactions on Microwave Theory and Techniques*, 11:869–81.

[Hejazi et al., 1998] Hejazi, Z., Excell, P., and Jiang, Z. (1998). Accurate distributed inductance of spiral resonators. *IEEE Microwave and Guided Wave Letters*, 8:164–6.

[Huan-Shang et al., 1997] Huan-Shang, T., Lin, L. J., Frye, R. C., Tai, K. L., Lau, M. Y., Kossives, D., Hrycenko, F., and Young-Kai, C. (1997). Investigation of current crowding effect on spiral inductors. In *IEEE MTT-S International Microwave Symposium Digest*, pages 139–142.

[Hurley and Duffy, 1995] Hurley, W. G. and Duffy, M. C. (1995). Calculation of self and mutual impedances in planar magnetic structures. *IEEE Trans. Magnetics*, 31:2416–22.

[Hurley and Duffy, 1997] Hurley, W. G. and Duffy, M. C. (1997). Calculation of self- and mutual impedances in planar sandwich inductors. *IEEE Trans. Magnetics*, 33:2282–90.

[Jackson, 1999] Jackson, J. D. (1999). *Classical electrodynamics*. Wiley, New York, 3rd edition.

[Jaeger, 1993] Jaeger, R. C. (1993). *Introduction to microelectronic fabrication*. Addison-Wesley Pub. Co., Reading, Mass.

[Jansen et al., 1997] Jansen, S., Negus, K., and Lee, D. (1997). Silicon bipolar vco family for 1.1 to 2.2 ghz with fully-integrated tank and tuning circuits. In *IEEE International Solid-State Circuits Conference*, pages 392–93.

[Jiang et al., 1997] Jiang, Z., Excell, P. S., and Hejazi, Z. M. (1997). Calculation of distributed capacitances of spiral resonators. *IEEE Transactions on Microwave Theory and Techniques*, 45:139–42.

[Johns and Martin, 1997] Johns, D. and Martin, K. W. (1997). *Analog integrated circuit design*. Wiley, New York.

[Johnson et al., 1996] Johnson, R. A., Chang, C. E., Asbeck, P. M., Wood, M. E., Garcia, G. A., and Lagnado, I. (1996). Comparison of microwave inductors fabricated on silicon-on-sapphire and bulk silicon. *IEEE Microwave and Guided Wave Letters*, 6:323–5.

[Kaertner, 1990] Kaertner, F. X. (1990). Analysis of white and $f^{-\alpha}$ noise in oscillators. *Int. J. Circuit Theory Appl.*, 18:485–519.

[Kamogawa et al., 1999] Kamogawa, K., Nishikawa, K., Toyoda, I., Tokumitsu, T., and Tanaka, M. (1999). A novel high-q and wide-frequency-range inductor using si 3-d mmic technology. *IEEE Microwave and Guided Wave Letters*, 9:16–18.

[Kamon et al., 1994a] Kamon, M., Ttsuk, M. J., and White, J. K. (1994a). Fasthenry: a multipole-accelerated 3-d inductance extraction program. *IEEE Transactions on Microwave Theory and Techniques*, 42:1750–8.

[Kamon et al., 1994b] Kamon, M., Ttsuk, M. J., and White, J. K. (1994b). Fasthenry: a multipole-accelerated 3-d inductance extraction program. *IEEE Trans. Microwave Theory Tech.*, 42:1750–58.

[Kapur and Long, 1997] Kapur, S. and Long, D. E. (1997). Ies3: a fast integral equation solver for efficient 3-dimensional extraction. In *IEEE/ACM International Conference on Computer-Aided Design*, pages 448–55.

[Kim et al., 1995] Kim, B.-K., Ko, B.-K., Lee, K., Jeong, J.-W., Lee, K.-S., and Kim, S.-C. (1995). Monolithic planar rf inductor and waveguide structures on silicon with performance comparable to those in gaas mmic. In *International Electron Devices Meeting*, pages 717–20.

[Kim and O, 1997] Kim, K. and O, K. (1997). Characteristics of an integrated spiral inductor with an underlying n-well. *IEEE Transactions on Electron Devices*, 44:1565–7.

[Kittel, 1996] Kittel, C. (1996). *Introduction to solid state physics*. Wiley, New York, 7th edition.

[Kouznets and Meyer,] Kouznets, K. and Meyer, R. G. Phase noise in lc oscillators. submitted for publication.

[Krafcsik and Dawson, 1986] Krafcsik, D. M. and Dawson, D. E. (1986). A closed-form expression for representing the distributed nature of the spiral inductor. In *IEEE Microwave and Millimeter-Wave Monolithic Circuits Symposium. Digest of Papers*, pages 87–92.

[Kuhn et al., 1995] Kuhn, W. B., Elshabini-Riad, A., and Stephenson, F. W. (1995). Centre-tapped spiral inductors for monolithic bandpass filters. *Electronics Letters*, 31:625–6.

[Lee et al., 1998a] Lee, J., Karl, A., Abidi, A. A., and Alexopoulos, N. G. (1998a). Design of spiral inductors on silicon substrates with a fast simulator. In *European Solid-State Integrated Circuits Conference*.

[Lee et al., 1998b] Lee, Y.-G., Yun, S.-K., and Lee, H.-Y. (1998b). Novel high-q bondwire inductor for mmic. In *International Electron Devices Meeting*, pages 548–51.

[Leeson, 1966] Leeson, D. B. (1966). A simple model for oscillator noise spectrum. *Proc. IEEE*, 54:329–330.

[Long and Copeland, 1997] Long, J. R. and Copeland, M. A. (1997). The modeling, characterization, and design of monolithic inductors for silicon rf ic's. *IEEE J. Solid-State Circuits*, 32:357–69.

[Long et al., 1996] Long, J. R., Copeland, M. A., Kovacic, S. J., Malhi, D. S., and Harame, D. L. (1996). Rf analog and digital circuits in sige technology. In *IEEE International Solid-State Circuits Conference*, pages 82–83.

[Lopez-Villegas et al., 2000] Lopez-Villegas, J. M., Samitier, J., Cane, C., Losantos, P., and Bausells, J. (2000). Improvement of the quality factor of rf integrated inductors by layout optimization. *IEEE Transactions on Microwave Theory and Techniques*, 48:76–83.

[Lovelace et al., 1994] Lovelace, D., Camilleri, N., and Kannell, G. (1994). Silicon mmic inductor modeling for high volume, low cost applications. *Microwave Journal*, 37:60–71.

[Mahmoud and Beyne, 1997] Mahmoud, S. F. and Beyne, E. (1997). Inductance and quality-factor evaluation of planar lumped inductors in a multilayer configuration. *IEEE Trans. Microwave Theory Tech.*, 45:918–23.

[Merrill et al., 1995] Merrill, R. B., Lee, T. W., You, H., Rasmussen, R., and Moberly, L. A. (1995). Optimization of high q integrated inductors for multi-level metal cmos. In *International Electron Devices Meeting*, pages 983–6.

[Meyer et al., 1997] Meyer, R. G., Mack, W. D., and Hageraats, J. J. E. M. (1997). A 2.5-ghz bicmos transceiver for wireless lans. *IEEE Journal of Solid-State Circuits*, 12:2097–104.

[Mohan et al., 1999] Mohan, S. S., del Mar Hershenson, M., Boyd, S. P., and Lee, T. H. (1999). Simple accurate expressions for planar spiral inductances. *IEEE Journal of Solid-State Circuits*, 34:1419–24.

[Mohan et al., 1998] Mohan, S. S., Yue, C. P., del Mar Hershenson, M., Wong, S. S., and Lee, T. H. (1998). Modeling and characterization of on-chip transformers. In *International Electron Devices Meeting*, pages 531–4.

[Muller and Kamins, 1986] Muller, R. S. and Kamins, T. I. (1986). *Device electronics for integrated circuits*. Wiley, New York, 2nd edition.

[Nabors and White, 1991] Nabors, K. and White, J. K. (1991). Fastcap: a multipole accelerated 3-d capacitance extraction program. *IEEE Transactions on Computer-Aided Design of Integrated Circuits and Systems*, 10:1447–59.

[Neudeck, 1983] Neudeck, G. W. (1983). *The bipolar junction transistor*. Addison-Wesley, Reading, Mass.

[Neudeck, 1989] Neudeck, G. W. (1989). *The PN junction diode*. Addison-Wesley, Reading, Mass., 2nd edition.

[Nguyen and Meyer, 1990] Nguyen, N. M. and Meyer, R. G. (1990). Si ic-compatible inductors and lc passive filters. *IEEE J. Solid-State Circuits*, 27:1028–31.

[Niknejad,] Niknejad, A. M. Discussion during ieee 1997 custom integrated circuits conference author interview.

[Niknejad, 1997] Niknejad, A. M. (1997). Analysis, design, and optimization of spiral inductors and transformers for si rf ics: research project. Master's thesis, University of California, Berkeley.

[Niknejad et al., 1998] Niknejad, A. M., Gharpurey, R., and Meyer, R. G. (1998). Numerically stable green function for modeling and analysis of substrate coupling in integrated circuits. *IEEE Trans. CAD*, 17:305–315.

[Niknejad and Meyer,] Niknejad, A. M. and Meyer, R. G. Asitic: Analysis of si inductors and transformers for ics. http://www.eecs.berkeley.edu/~niknejad.

[Niknejad and Meyer, 1997] Niknejad, A. M. and Meyer, R. G. (1997). Analysis and optimization of monolithic inductors and transformers for rf ics. In *IEEE International Solid-State Circuits Conference*, pages 375–8.

[Niknejad and Meyer, 1998] Niknejad, A. M. and Meyer, R. G. (1998). Analysis, design, and optimization of spiral inductors and transformers for si rf ics. *IEEE J. Solid-State Circuits*, 33:1470–81.

[Park et al., 1997a] Park, M., Kim, C. S., Park, J. M., Yu, H. K., and Nam, K. S. (1997a). High q microwave inductors in cmos double-metal technology and its substrate bias effects for 2 ghz rf ics application. In *International Electron Devices Meeting*, pages 59–62.

[Park et al., 1997b] Park, M., Lee, S., Yu, H. K., Koo, J. G., and Nam, K. S. (1997b). High q cmos-compatible microwave inductors using double-metal interconnection silicon technology. *IEEE Microwave and Guided Wave Letters*, 7:45–7.

[Percival, 1937] Percival, W. S. (1937). British Patent 460562.

[Pettenpaul and et al., 1988] Pettenpaul, E. and et al., H. K. (1988). Cad models of lumped elements on gaas up to 18 ghz. *IEEE Trans. Microwave Theory Tech.*, 36:294–304.

[Pierret, 1990] Pierret, R. F. (1990). *Field effect devices*. Addison-Wesley Pub. Co., Reading, Mass., 2nd ed edition.

[Poritsky and Jerrard, 1954] Poritsky, H. and Jerrard, R. P. (1954). Eddy-current losses in a semi-infinite solid due to a nearby alternating current. *Trans. Am. Inst. Elect. Engrs, part I*, 73:97–106.

[Pozar, 1997] Pozar, D. M. (1997). *Microwave Egineering*. Wiley, New York, NY, 2nd edition.

[Proakis, 1995] Proakis, J. G. (1995). *Digital communications*. McGraw-Hill, New York, 3rd edition.

[Rabjohn, 1991] Rabjohn, G. G. (1991). Monolithic microwave transformers. Master's thesis, Carleton University.

[Ramo et al., 1994] Ramo, S., Whinnery, J. R., and Duzer, T. V. (1994). *Fields and Waves in Communication Electronics*. John Wiley and Sons, Inc., New York, 3rd ed. edition.

[Rategh and Lee, 1999] Rategh, H. R. and Lee, T. H. (1999). Superharmonic injection-locked frequency dividers. *IEEE Journal of Solid-State Circuits*, 34:813–21.

[Rautio, 1999] Rautio, J. C. (1999). Free em software analyzes spiral inductor on silicon. *Microwaves and RF*, pages 165–9.

[Razavi, 1998] Razavi, B. (1998). *RF microelectronics*. Prentice Hall, Upper Saddle River, NJ.

[Rejaei et al., 1998] Rejaei, B., Tauritz, J. L., and Snoeij, P. (1998). A predictive model for si-based circular spiral inductors. In *IEEE MTT-S International Microwave Symposium Digest*, pages 148–54.

[Roach, 1982] Roach, G. F. (1982). *Green's Functions*. Cambridge University Press, Cambridge; New York, 2nd edition.

[Rofougaran et al., 1998a] Rofougaran, A., Chang, G., Rael, J. J., Chang, J. Y.-C., Rofougaran, M., Chang, P., Djafari, M., Ku, M.-K., Roth, E., Abidi, A., and Samueli, H. (April 1998a). A single-chip 900-mhz spread-spectrum wireless transceiver in 1-μm cmos. i. architecture and transmitter design. *IEEE Journal of Solid-State Circuits*, 33(4):515–34.

[Rofougaran et al., 1998b] Rofougaran, A., Chang, G., Rael, J. J., Chang, J. Y.-C., Rofougaran, M., Chang, P., Djafari, M., Min, J., Roth, E., Abidi, A., and Samueli, H. (April 1998b). A single-chip 900-mhz spread-spectrum wireless transceiver in 1-μm cmos. ii. receiver design. *IEEE Journal of Solid-State Circuits*, 33(4):535–47.

[Rudell et al., 1997] Rudell, J., Ou, J.-J., Cho, T., Chien, G., Brianti, F., Weldon, J., and Gray, P. R. (1997). A 1.9 ghz wide-band if double conversion cmos integrated receiver for cordless telephone applications. In *IEEE International Solid-State Circuits Conference*, pages 304–5, 476.

[Ruehli, 1974] Ruehli, A. (1974). Equivalent circuit models for three-dimensional multiconductor systems. *IEEE Transactions on Microwave Theory and Techniques*, MTT-22:216–21.

[Ruehli, 1972] Ruehli, A. E. (1972). Inductance calculations in a complex integrated circuit environment. *IBM J. Res. Develop.*, 16:470–481.

[Ruehli and Heeb, 1992] Ruehli, A. E. and Heeb, H. (1992). Circuit models for three-dimensional geometries including dielectrics. *IEEE Trans. Microwave Theory Tech*, 40:1507–1516.

[Ruehli et al., 1995] Ruehli, A. E., Miekkala, U., and Heeb, H. (1995). Stability of discretized partial element equivalent efie circuit models. *IEEE Trans. Antennas Propag.*, 43:553–9.

[Samori et al., 1998] Samori, C., Lacaita, A. L., Villa, F., and Zappa, F. (1998). Spectrum folding and phase noise in lc tuned oscillators. *IEEE Trans. Circuits Syst. II, Analog Digit. Signal Process*, 45:781–90.

[Schmuckle, 1993] Schmuckle, F. J. (1993). The method of lines for the analysis of rectangular spiral inductors (in mmics). *IEEE Transactions on Microwave Theory and Techniques*, 41:1183–6.

[Shepherd, 1986] Shepherd, P. R. (1986). Analysis of square-spiral inductors for use in mmic's. *IEEE Trans. on MTT*, MTT-34:467–472.

[Soyuer et al., 1995] Soyuer, M., Burghartz, J. N., Jenkins, K. A., Ponnapalli, S., Ewen, J. F., and Pence, W. E. (1995). Multilevel monolithic inductors in silicon technology. *Electronics Letters*, 31:359–60.

[Steyaert et al., 1998] Steyaert, M., Borremans, M., Janssens, J., de Muer, B., Itoh, I., Craninckx, J., Crols, J., Morifuji, E., Momose, S., and Sansen, W. (1998). A single-chip cmos transceiver for dcs-1800 wireless communications. In *IEEE International Solid-State Circuits Conference*, pages 48–9, 411.

[Stoll, 1974] Stoll, R. L. (1974). *The Analysis of Eddy Currents*. Clarendon Press, Oxford.

[Tegopoulos and Kriezis, 1985] Tegopoulos, J. A. and Kriezis, E. E. (1985). *Eddy Currents in Linear Conducting Media*. Elsevier Science Pub. Co., Amsterdam; New York: Elsevier; New York: Distributors for the U.S. and Canada.

[Trim, 1990] Trim, D. W. (1990). *Applied Partial Differential Equations*. PWS-KENT, Boston.

[Vendelin and A. M. Pavio, 1990] Vendelin, G. D. and A. M. Pavio, U. L. R. (1990). *Microwave Circuit Design Using Linear and Nonlinear Techniques*. Wiley, New York, NY.

[Wait, 1982] Wait, J. R. (1982). *Geo-electromagnetism*. Academic Press, New York.

[Weeks et al., 1979] Weeks, W. T., Wu, L. L., McAllister, M. F., and Singh, A. (1979). Resistive and inductive skin effect in rectangular conductors. *IBM J. Res. Develop.*, 23:652–660.

[Wolfram, 1999] Wolfram, S. (1999). *The Mathematica Book*. Wolfram Media; Cambridge University Press, Champaign, IL; New York, 4th edition.

[Yoon et al., 1998] Yoon, J.-B., Kim, B.-K., Han, C.-H., Yoon, E., Lee, K., and Kim, C.-K. (1998). High-performance electroplated solenoid-type integrated inductor si^2 for rf applications using simple 3d surface micromachining technology. In *International Electron Devices Meeting*.

[Yoshitomi et al., 9998] Yoshitomi, T., Sugawara, Y., Morifuji, E., Ohguro, T., Kimijima, H., Morimoto, T., Momose, H. S., Katsumata, Y., and Iwai, H. (19998). On-chip spiral inductors with diffused shields using channel-stop implant. In *International Electron Devices Meeting*, pages 540–3.

[Young and Boser, 1197] Young, D. J. and Boser, B. E. (1197). A micromachine-based rf low-noise voltage-controlled oscillator. In *IEEE International Solid-State Circuits Conference*, pages 431–4.

[Young et al., 1997] Young, D. J., Malba, V., Ou, J.-J., Bernhardt, A. F., and Boser, B. E. (1997). A low-noise rf voltage-controlled oscillator using on-chip high-q three-dimensional coil inductor and micromachined variable capacitor. In *International Electron Devices Meeting*, pages 128–31.

[Yue and Wong, 1997] Yue, C. and Wong, S. (1997). On-chip spiral inductors with patterned ground shields for si-based rf ic's. In *IEEE Symposium on VLSI Circuits*, pages 85–6.

[Yue and Wong, 1998] Yue, C. P. and Wong, S. S. (1998). On-chip spiral inductors with patterned ground shields for si-based rf ic's. *IEEE Journal of Solid-State Circuits*, 13:743–52.

[Zannoth et al., 1998] Zannoth, M., Kolb, B., Fenk, J., and Weigel, R. (1998). A fully integrated vco at 2 ghz. In *IEEE International Solid-State Circuits Conference*, pages 224–5.

[Zhou, 1993] Zhou, P.-b. (1993). *Numerical analysis of electromagnetic fields*. Springer-Verlag, Berlin ; New York.

Index

ASITIC, 97–107
    capacitors, 133
    geometry engine, 99
    library dependence, 99
    numerical calculations, 100–101
    organization, 99–100
    technology file, 99
    visualization, 106–107

ABCD matrix, 150
active area, 134
active device
    typical examples, 12
aluminum, 16, 17, 88
    sheet resistance, 17
    spiking, 17
    thickness, 17
analysis
    capacitors, 103
    transformers, 103–106
    two-port, 102–103
anisotropy, 46
antenna, 6, 15
    efficiency, 126
applications
    artificial transmission line, 5, *see* distributed amplifier
    balun, 5
    center-tapped transformer, 5
    differential operation, 5
    distributed amplifier, 5
    filters, 5
    impedance matching, 4
    LC tuned load, 5
    low noise amplifier, 4, 7
    mixer, 4
    phase locked loops, 7
    power amplifier, 4
    series-feedback, 5
    tank, 7
    traveling-wave amplifier, 5
    voltage controlled oscillator, 7
    wireless transceiver, 6
arithmetic mean distance (AMD), 68
attenuation constant, 150, 159, 160, 169
avalanche
    breakdown, 32
    multiplication, 36

balun, 28
base-pinch, 34
basis function
    spatially localized, 54
battery life, 8
Bessel function, 93
bias voltage, 141
BiCMOS, 8–9, 20, 21, 34, 75, 109, 128
bipolar, 20, 21, 34, 75, 109, 128, 139
bisected-π m-derived section, 155
black-box, 11
BLAS, 99, 100
bond
    pad, 29, 35
    pad capacitance, 111
    wire, 21, 26, 42
bottom plate, 134
boundary conditions, 21, 50, 54, 79, 80, 83, 103, 168, 169
boundary method, 52
breakdown, 32
broadband, 5, 149
bulk substrate, 21, 75
buried oxide isolation, 42
bypass, 132

calibration, 110
capacitance, 59
    pn-junction, 33
    linear, 132–133
    matrix, 57, 59, 104

184 INDUCTORS AND TRANSFORMERS FOR SI RF ICS

tolerance, 35
capacitive
    coupling, 28, 102
    current injection, 129
capacitor, 34–35
    distributed, 167–169
    metal-metal, 34, 133–134, 141
    MOS, 33–34, 133
    poly-poly metal, 34
cellular communication, 6
center-tap, 22
center-tapped
    inductor, 129
    transformers, 28
channel length, 158
charge and current discretization, 53
charge conservation, 100
chemical vapor deposition, 20
circular, 22
circular symmetry, 43, 81
CMOS, 8–9, 16, 17, 20, 21, 34, 42, 75, 109, 112, 128, 152, 160
co-planar probes, 110
coil, 22, 39, 41
Colpitts oscillator, 140
common ground, 103
common mode rejection, 135
complex power, 60
conduction current, 31
conductive
    losses, 48
    substrate, 18–21
conservation of energy, 13
constitutive relations, 45
constructive interference, 150
copper, 17, 88
coupled inductors, 109
coupled transmission lines, 40, 129, 149
coupling factor, 28
current constriction, 27, 43, 72, 106
current crowding, 17–18, 43
current density plot, 73, 100
current distribution, 17
    high frequency, 68
cutoff frequency, 151, 159, 160

depletion, 33
depletion mode, 34
depletion region, 31
    thickness, 32, 34
device layout, 21–31
diamagnetic, 19, 46, 76
die photo, 144
dielectric
    effective, 49
dielectric constant, 40
dielectric losses, 152

differential, 5
    circuits, 22
    operation, 129
    quad, 141
    quality factor, 129
    structure, 134
diffusion, 20
    current, 31
    length, 32
    process, 9
    resistor, 34
digital
    high speed, 17, 59, 163, 164
    integration, 8
    interconnect, 164
    market growth, 9
    noise, 8, 36, 164
digital signal processing, 7, 8
diode, 12, 31–33, 133
Dirac delta function, 51
direct conversion, 7
discrete components, 6
discrete cosine transform, 100
discretization
    charge, 55
    current, 54
displacement current, 18, 32, 35, 53, 65, 88, 90, 104
dissipation matrix, 12
distributed amplifier
    actively loaded line, 160–161
    artificial transmission line, 150
    design tradeoffs, 158–160
    gain, 156–158
    introduction, 149–150
    number of stages, 150
    optimal number of stages, 159–160
distributed effects, 40
distributed resistance, 34
divider, 134
    current consumption, 143
    headroom, 143
    latch based, 143
    speed, 143
domain techniques, 52
doping, 141
    profile, 52
down-bonds, 21
down-conversion, 146
drain, 36
    line, 5, 149, 155
    line impedance, 159
    lossy line, 151
dynamic range, 5, 8

eddy currents, 9, 21, 26, 29, 41, 43, 72, 76, 77, 80, 81, 102, 109, 113, 122, 152
    3-d solution, 81–84

Index 185

boundary value problem, 78–81
circular symmetry, 81
combating, 26
equation, 50
generated by square spirals, 92–93
high frequency, 88–92
losses, 75, 91
low frequency, 84–88
single substrate layer, 93
substrate, 18
two layer substrate, 94
efficiency, 5, 8
eigenfunction, 54
electric
center, 23
energy storage, 32
force, 69
response, 76
electrically short open-line, 167
electromigration, 17
EM simulation, 42
emitter
coupled logic, 134
follower, 134, 143
energy storage, 11, 14
entropy, 15
epitaxial growth, 109
epitaxial layer, 21
epitaxial pinch, 34
epitaxy, 20
epoxy, 21
equivalent
resistance, 85
equivalent circuit, 129
near resonance, 119
shielded MIM capacitors, 133
external components, 5

Faraday's law, 60, 62, 69
FASTHENRY, 89
feedback
factor, 134
network, 141
ferromagnetic, 46
FET
losses, 151
model, 153, 156
FFTW, 99
filament, 39
filters
channel selection, 126
gm-C, 5
MOSFET-C, 5, 34
surface acoustic wave, 5
finite difference equation, 52
finite difference time domain, 43
finite element method, 52

flicker noise, 138–140
free-space, 50, 80, 90
kernel, 89
frequency division, 127, 142–143
frequency division multiplexing, 126
frequency hopped multiple access, 126

G-S-G probes, 111–113
GaAs, 8–9, 18, 40, 48
gain compression, 31
gain-bandwidth product, 149
gate
line, 5, 149, 155
line impedance, 159
lossy line, 151
oxide, 33
resistance, 158
gauge
Coulomb, 46, 76, 78
Lorenz, 46, 48
lossy media, 48
geometric
center, 23
gold, 17, 40, 88
Green function, 43, 50–53, 87, 88, 100, 101, 103, 106
averaged, 65
circular, 43
dyadic, 51–52, 63
electric, 53
free-space, 43, 75
full-wave, 43
magnetic, 54
Green's identities, 51, 52
Greenhouse, 39, 40, 42
ground, 21, 111
back-plane, 18, 21, 35
bounce, 31
conductors, 102
inductance, 31
on-chip, 21
plane, 68
secondary, 104
structure, 29
substrate, 41
Grover, 39, 42
gyrator, 14

half-wavelength, 129
heat, 15, 17, 18, 128
Helmholtz
equations, 52
theorem, 47
Hermitian matrix, 13
high quality factor, 7
high-field region, 36
hollow spiral, 27, 43
hot electron effects, 36

Hypergeometric function, 93
hysteresis, 46

image
    currents, 20, 29, 40, 83, 85, 91, 113
    impedance, 153
    impedance matching, 155
    parameter method, 150
        lossless transmission line, 151
        lossy transmission line, 151–153
impedance
    match, 4, 5, 17
impulse response function, 51
incident power, 12
induced
    electric field, 60
    magnetic losses, 75
    substrate currents, 18
inductance
    bond wire, 4
    calculation
        conductors, 67
        filamental, 66–67
        geometric mean distance, 39, 67–68, 90
        high frequency, 68–72
        hybrid method, 91
        parallel connection, 65–66
        series connection, 65
    curve-fit, 43
    definition, 60–65
        energy, 60–61
        magnetic flux, 61–62
        magnetic vector potential, 62–65
    external, 70
    frequency-independent, 83
    high frequency, 68–72
    internal, 60
    lead frame, 4
    low frequency, 64
    mutual, 62, 63, 66, 90
    reflected, 83, 102
    self, 17, 62, 75
    simulated, 4
    tolerance, 26, 42
inductor
    layout
        non-planar, 24–26
        planar, 22–24
    losses, 151
    multi-layer, 44
    polygon spiral, 109
    series connected, 41, 44
    shunt connected, 41
    stacked, 44
injection lock, 142
inner turns of spiral, 72

instability, 31, 36
integral operator
    kernel, 87
integrated circuit, 15, 17, 24, 31, 55, 59
    cross-section of metal layers, 16–18
    radio frequency, 21
    technology, 8
interconnect, 17, 35
intermediate frequency, 125, 126
interwinding capacitance, 20, 26, 119
inverse cosine transform, 80
inversion, 34
ion implantation, 20
irreversible process, 15
irrotational, 47
isolation, 31, 113, 132
iterative solution, 89

jamming, 31

k-factor, 28
Kirchhoff's current law, 56, 89, 100, 102
Kirchhoff's voltage law, 100, 102, 136
Krylov sub-space iteration, 43

LAPACK, 99, 100
latch, 143
latch-up, 21, 109
lateral
    coil, 26, 41
    magnetic fields, 26
Leeson formula, 128
Lenz's law, 60, 85
linear superposition, 51
linearity, 5, 12, 46, 76, 87
LINPACK, 99
lithographic technology, 9
local oscillator, 126, 142, 146
logarithmic singularity, 83
loop
    gain, 138, 140
    inductor, 39
Lorentzian distribution, 15
loss tangent, 48, 78, 103
lossy capacitor, 30

m-derived matching network, 159
magnetic
    coupling, 102, 134
    energy, 60
    flux, 61
    force, 69
    monopole, 59
    mutual coupling, 22, 25
    response, 76
    vector potential, 62–63, 67
majority carriers, 33
Manhattan geometry, 42

manufacturability
	non-planar structure, 119
master-slave, 143
matching, 141, 143
material properties, 15
matrix compression, 102
matrix-fill
	capacitance, 100
	inductance, 101
	operation, 100–101
Maxwell, 39
Maxwell's equations, 39, 57, 76, 87, 88
	discretization, 53–57
	inversion, 50–52
	numerical solution, 52–57
	time-period, 45
measurement, 109–122
	s-parameters, 113, 119
	device layout, 110
	eddy currents, 116
	effective inductance, 113
	effective series resistance, 113
	multi-layer structure, 119–122
	negative resistance, 116
	quality factor, 116, 120
	self-resonant frequency, 116, 119
	setup and calibration, 110–113
	single layer structure, 113–116
metal
	layers, 16
	losses, 16–18, 27, 40
		eddy currents, 17–18
		ohmic, 17
	pitch, 28
	spacing, 21, 27, 70, 72
method of moments, 43, 53
micro-machined solenoid inductors, 42
microstrip, 150
minority carriers, 32, 34
mixed-signal ICs, 36
mixer, 125, 126, 142
mode-locking, 31
modulation, 126
MOS, 9
multi-conductor system, 18
multi-layer
	inductor
		series-connected, 110
	spirals
		series-connected, 103
		shunt-connected, 103
	substrate, 76
multi-level metallization, 41

natural frequency, 13
negative feedback, 31
negative resistance, 13, 134, 160

Neumann's equation, 64
NMOS, 8
noise, 5
	control line, 131
	immunity, 134
	injection, 5
	power spectral density, 138
	up-conversion, 140
non-linear, 5, 12, 46
non-uniform
	current distribution, 27, 43, 72, 75, 91, 106
	media, 52
numerical integration, 54, 101
	magnetic vector potential, 90–92
numerical solution
	assumptions, 53
	current flow, 88

off-chip, 5, 6, 31
ohmic losses, 75, 103
on-chip, 5, 6, 31, 60
	filter, 142
OpenGL, 100
optimization, 72, 155
optimum bias current, 141
orthonormal set, 54
oscillation amplitude, 130
output buffers, 143
oxide, 17, 20, 103
	thickness, 33, 34, 119, 133

package, 4, 8, 21
	paddle, 21
palladium, 17
panels
	constant charge, 101
paramagnetic, 46
parametric amplifier, 12
parasitic
	coupling, 8, 31
	magnetic coupling, 129
partial
	inductance, 69
	inductance matrix, 25, 42, 54, 57, 65, 70, 89, 102
		reduction, 89
passive
	area, 134
	network, 13
passive device
	applications, 4–8
	definition, 11–13
	high power, 17
	metal losses, 16–18
	on-board, 3
	on-chip, 3
	quality factor, 14–15
	reciprocity, 13–14

stability, 13
substrate coupling, 36
substrate losses, 18–19
typical examples, 11
passivity, 13
PEEC, 39, 43, 57, 75, 88, 90, 100–106
penetration depth, 17, 21
perfectly coupled inductors, 26
phase
    noise, 7, 126, 127, 142, 146
    velocity, 150, 157
planar metallization, 21
planar structure, 120
planarization constraints, 20
platinum, 17
pn junction
    abrupt, 32
    barrier, 32
    built-in field, 32
    forward biased, 32, 109, 130
    leakage currents, 32
    linearly graded, 33
    reverse biased, 31, 32, 34, 41, 129, 141
    reverse-based, 36
    space-charge region, 32
Poisson's equation, 51, 53, 55, 60, 78
    modified, 49
    vector form, 62
polygon, 22
polyimide, 20, 41
polysilicon, 16, 28, 41, 42
positive definite matrix, 13
positive feedback, 31, 134
post-processing, 21
potential
    electric scalar, 46–47
    electromagnetic, 48
    magnetic vector, 46–47
power
    amplifier, 5, 127
    delivered to one-port, 14
    dissipation, 14
    gain, 12
    injection, 5
Poynting's theorem, 48, 84
previous work
    early research, 39
    GaAs substrate, 40
    Si substrate, 40–44
process parameters, 110, 119
process variation, 141
propagation
    constant, 150–151, 153, 169
    mode, 19
    modeg, 19
    velocity, 157
proximity, 16, 20, 29
    effect, 17–18, 27, 42, 75, 103, 152
punctured matrix, 104

QUADPACK, 99, 101
quality factor, 14–15, 25, 26, 29, 33, 34, 44, 126
    capacitor, 122
    high frequency, 26, 127, 129
    metal losses, 16–18
    substrate losses, 18–19
quarter-wavelength, 129
quasi-static, 76

radiation, 18, 47, 127, 128
radio frequency, 29
reactance
    capacitive, 14
    frequency variation, 60
    inductive, 14
reactive energy, 16
reciprocal mixing, 127
reciprocity, 13–14, 64
recursive-SVD
    algorithm, 43
reflected power, 12
reflection coefficient, 13
resistance
    effective, 167
    high frequency, 18, 68, 70–72
resistivity, 21
resistor, 7, 34–35
    MOS, 34
    polysilicon, 34
    well, 34
reversible process, 15
ring inductor, 39
rotational
    invariance, 81
    symmetry, 43
Ruehli, 39

S-G probes, 111–113
scattering matrix, 12
    symmetric, 13
selective etching, 18, 40
self
    adjoint, 54
    resonance, 26
        interwinding capacitance, 119
        substrate capacitance, 119
    resonant frequency, 9, 17, 20, 27, 30–31, 40, 41, 129
        measured, 113
separation of variables, 78, 82
series connection, 24
series to parallel transformation, 30
shield, 28–31, 41, 113, 122, 133
    currents, 29
    shallow junction diffused, 42

Index

short-channel MOS transistor, 36
shot noise, 134
shunt connection, 24
Si, 8–9, 18, 19, 21, 46, 48, 76, 133
    compatible inductors, 40
    intrinsic, 20, 41
    substrate resistivity, 18
SiGe, 8–9
signal-to-noise ratio, 36
single-ended, 5, 129
singular value decomposition, 90
skin depth, 21, 59, 101
skin effect, 17–18, 27, 39, 42, 72, 75, 103, 113, 152
    bulk, 78
sliver, 17
SNR, 31, 127
solenoidal, 18, 47, 62, 64
solver
    iterative, 100
SONNET, 44
space-charge, 31
spanning set, 54
spectral leakage, 7, 31, 127
SpectreRF, 136, 139, 143
spiral, 22, 39
    inductor, 18
spurs, 31
square, 22
stability, 13, 160
    conditional, 13
staggered inductors, 41
steady-state oscillation, 138
stochastic bandpass signal, 126
Stoke's theorem, 62
stratified substrate, 20, 55
substrate, 8
    conductive, 21, 128
    conductivity, 20
    contact, 35
        halo, 31, 41
    coupling, 29, 31, 35–36, 111, 134
    current injection, 35
    currents, 91
    discretization, 53
    epi, 9, 43, 128, 152
    heavily resistive, 40
    high resistivity, 41
    induced losses, 21
    injection, 133
    insulating, 8, 18, 41, 152
    isolation, 31
    layers, 20
    loss matrix
        algorithm, 93
    losses, 9, 18–19, 26, 27, 48, 53, 88
        calculation of, 48–50

    curtailing, 129
    electrically induced, 102, 103
    magnetically induced, 103
    profile, 20–21
        bipolar, 20–21
        CMOS, 20–21
    removal, 18
    thickness, 21
superheterodyne receiver, 7, 126
surface waves, 19
swing, 141
symmetric inductors, 22

T-coils, 160
T-section, 151, 152
    center, 157
tank
    loading, 31
    quality factor, 128, 139
tapered spirals, 27, 43
Taylor series, 169
temperature variation, 141
thermal noise, 134
thermionic emission, 20
thick metal, 17
thick oxide, 17, 20, 40
three-port model, 129
time-varying, 12, 131, 139
    magnetic field, 18, 27
    noise, 134
    transfer function, 136
titanium, 17
tolerance
    capacitance, 133
top-metal, 17
toroid, 22, 39
transceiver
    architecture, 7
    superheterodyne, 125
transformer, 18, 28, 43, 134
    k-factor, 28
    capacitive, 134
    center tap, 28
    metal-metal, 28
    non-planar, 109
    planar, 23, 109
    primary, 28, 103, 104
    secondary, 28, 103, 104
    turns ratio, 28
transistor, 12
    parasitics, 149
transmission coefficient, 13
tungsten, 17
tuning range, 33
tunneling, 32
twin-well, 21
two-port parameters, 101

uniform
    convergence, 54
    current distribution, 42
    elements, 53

vacuum tube, 12
varactor, 31–33, 129, 146
    high frequency, 130
    losses, 129–132
    quality factor, 33, 34, 141
    resonance, 132
vertical coil, 26
via, 23, 26, 55, 113
virtual ground, 134, 135
voltage controlled oscillator, 4, 33, 142, 146
    die photo, 144
    differential, 129
    differential q-factor, 129
    effective capacitance, 137
    implementation, 141–142
    inductor design, 128–129
    introduction, 125–127
    measurement, 144–146
    motivation, 127
    phase noise, 134–140
    simulation, 139
    test setup, 145–146
    topology, 134
    total capacitance, 137
    tuning range, 133
    varactor losses, 129–132
voltage headroom, 5

wave
    attenuation, 150
    propagation, 5, 18
wavelength, 19
weak inversion, 33, 34
well, 21
winding capacitance, 113
wireless communication, 17
wireless LAN, 127

y-parameters, 110

zero IF, 7